Submicron Multiphase Materials

MATERIALS RESEARCH SOCIETY SYMPOSIUM PROCEEDINGS VOLUME 274

Submicron Multiphase Materials

Symposium held April 28-30, 1992, San Francisco, California, U.S.A.

EDITORS:

Ronald H. Baney
Dow-Corning Corporation
Midland, Michigan, U.S.A.

Laura R. Gilliom
Sandia National Laboratories
Albuquerque, New Mexico, U.S.A.

Shin-Ichi Hirano
Nagoya University
Nagoya, Japan

Helmut K. Schmidt
Universität de Saarlandes
Saarbrücken, Germany

MRS

MATERIALS RESEARCH SOCIETY
Pittsburgh, Pennsylvania

Single article reprints from this publication are available through University Microfilms Inc., 300 North Zeeb Road, Ann Arbor, Michigan 48106

CODEN: MRSPDH

Copyright 1992 by Materials Research Society.
All rights reserved.

This book has been registered with Copyright Clearance Center, Inc. For further information, please contact the Copyright Clearance Center, Salem, Massachusetts.

Published by:

Materials Research Society
9800 McKnight Road
Pittsburgh, Pennsylvania 15237
Telephone (412) 367-3003
Fax (412) 367-4373

Library of Congress Cataloging in Publication Data

Submicron multiphase materials: symposium held April 28-30, 1992, San Francisco, California, U.S.A. / editors, I. Baney, R., II. Gilliom, L., III. Herano, S.-I., IV. Schmidt, H.
   p.   cm. -- (Materials Research Society symposium proceedings, ISSN 0272-9172; v. 274)
   Includes bibliographical references and index.
   ISBN 1-55899-169-7
   1. Composite materials--Congresses. 2. Microstructure--Congresses.
I. Baney, R. II. Gilliom, L. III. Herano, S.-I. IV. Schmidt, H.
V. Series: Materials Research Society symposium proceedings; v. 274.
TA418.9.C6S79   1992   92-27698
620.1'18--dc20   CIP

Contents

PREFACE vii

MATERIALS RESEARCH SOCIETY SYMPOSIUM PROCEEDINGS viii

PART I: ORGANIC/ORGANIC COMPOSITES

FRACTURE TOUGHNESS AND FRACTURE MECHANISMS OF PBT/PC/IM BLEND 3
Jingshen Wu, Yiu-Wing Mai, and Brian Cotterell

A NOVEL EPOXY RESIN TOUGHENED BY HTBN LIQUID RUBBER 11
Xiaozu Han, Zhankui Yun, and Fenchun Guo

STRUCTURE-PROPERTY RELATIONSHIPS IN THE TOUGHENING OF POLY(METHYL METHACRYLATE) BY SUB-MICRON SIZE, MULTIPLE-LAYER PARTICLES 17
A.C. Archer, P.A. Lovell, J. McDonald, M.N. Sherratt, and R.J. Young

A STUDY ON THE BLENDS OF EPOXY/POLYBUTADIENE AND THE APPLICATION TO THE ENCAPSULATION OF A CAPACITOR 25
Zhongyuan Ren and Liying Qui

IN-SITU PHASE SEPARATION OF AN AMINE-TERMINATED SILOXANE IN EPOXY MATRICES 31
D.F. Bergstrom, G.T. Burns, G.T. Decker, R.L. Durall, D. Fryrear, G.A. Gornowicz, M. Tokunoh, and N. Odagiri

HIGH TEMPERATURE POLYMER NANOFOAMS 37
J. Hedrick, J. Labadie, T. Russell, V. Wakharkar, and D. Hofer

LASER OR FLOOD EXPOSURE GENERATED ELECTRICALLY CONDUCTING PATTERNS IN POLYMERS 47
Joachim Bargon and Reinhard Baumann

CHARACTERIZATION OF THERMOTROPIC LIQUID CRYSTALLINE POLYMER BLENDS BY POSITRON ANNIHILATION LIFETIME SPECTROSCOPY 53
Robert A. Naslund, Phillip L. Jones, and Andrew Crowson

SMALL ANGLE NEUTRON SCATTERING STUDIES OF SINGLE PHASE IPNS 59
Barry J. Bauer, Robert M. Briber, Shawn Malone, and Claude Cohen

PART II: ORGANIC/INORGANIC COMPOSITES

HIGH GLASS CONTENT NON-SHRINKING SOL-GEL COMPOSITES VIA SILICIC ACID ESTERS 67
Mark W. Ellsworth and Bruce M. Novak

REINFORCEMENT FROM IN-SITU PRECIPITATED SILICA IN POLY-SILOXANE ELASTOMERS UNDER VARIOUS TYPES OF DEFORMATION ... 77
James E. Mark, Shuhong Wang, Ping Xu, and Jianye Wen

MOLECULAR WEIGHT DEPENDENCE OF DOMAIN STRUCTURE IN SILICA-SILOXANE MOLECULAR COMPOSITES 85
  Tamara A. Ulibarri, Greg Beaucage, Dale W. Schaefer, Bernard J. Olivier, and Roger A. Assink

REINFORCEMENT IN SILICONE ELASTOMERS. A SHORT REVIEW 91
  John C. Saam

INTERPENETRATING ORGANOMETALLIC POLYMER NETWORKS (IOPN's): NOVEL ADVANCED MATERIALS 103
  B. Corain, M. Zecca, C. Corvaja, G. Palma, S. Lora, and K. Jerabek

SYNTHESIS, CHARACTERIZATION, AND DYNAMICS OF A ROD/SPHERE ORGANOCERAMIC COMPOSITE LIQUID 109
  Mark A. Tracy and R. Pecora

SILICA/SILICONE NANOCOMPOSITE FILMS: A NEW CONCEPT IN CORROSION PROTECTION 115
  Theresa E. Gentle and Ronald H. Baney

SOL-GEL SYNTHESIS OF CERAMIC-ORGANIC NANO COMPOSITES 121
  Helmut K. Schmidt

VARIABLE FREQUENCY CONDUCTIVITY OF LAYERED POLYPYRROLE/V_2O_5 COMPOSITES 133
  D.C. DeGroot, J.L. Schindler, C.R. Kannewurf, Y.-J. Liu, C.-G. Wu, and M.G. Kanatzidis

PART III: INORGANIC/INORGANIC COMPOSITES

MICROSTRUCTURE DEVELOPMENT AND MECHANICAL PROPERTIES OF Ce-TZP/La-β-ALUMINA COMPOSITES 141
  Takashi Fujii, Hironobu Muragaki, Hiraku Hatano, and Shin-Ichi Hirano

Al_2O_3-ZrO_2 CERAMICS WITH SUBMICRON MICROSTRUCTURES OBTAINED THROUGH MICROWAVE SINTERING, PLASMA SINTERING AND SHOCK COMPACTION 149
  J. McKittrick, B. Tunaboylu, J.D. Katz, and W. Nellis'

MICROCRYSTALLINE CERAMIC COMPOSITES BY ACTIVE FILLER CONTROLLED REACTION PYROLYSIS OF POLYMERS 155
  Peter Greil, Michael Seibold, and Tobias Erny

SYNTHESIS OF CARBON/FERRITE COMPOSITE BY IN-SITU PRESSURE PYROLYSIS OF ORGANOMETALLIC POLYMERS 167
  Shin-Ichi Hirano, Toshinobu Yogo, Koichi Kikuta, and Makoto Fukuda

MAGNETIC-PROPERTIES OF MECHANICALLY ALLOYED NANO-CRYSTALLINE Cu/Fe COMPOSITES 177
  C.P. Reed, S.C. Axtell, R.J. De Angelis, B.W. Robertson, V.V. Munteanu, and S.H. Liou

AUTHOR INDEX 183

SUBJECT INDEX 185

Preface

This book contains the papers presented at Symposium R of the Spring 1992 Materials Research Society meeting held in San Francisco, California. The title of the symposium, Submicron Multiphase Materials, was selected by the organizers to encompass the realm of composite materials from those smaller than conventional fiber matrix composites to those with phase separation dimensions approaching molecular dimensions.

The development of composite materials is as old as the development of materials. Humans quickly learned that, by combining materials, the best properties of each can be realized and that, in fact, synergistic effects often arise. For example, chopped straw was used by the Israelites to limit cracking in bricks. The famed Japanese samurai swords were multilayers of hard oxide and tough ductile materials. One also finds in nature examples of composite materials. These range from bone to wood, consisting of a hard phase which provides strength and stiffness and a softer phase for toughness.

Advanced composites are generally thought of as those which are based on a high modulus, discontinuous, chopped or woven fiber phase and a continuous polymer phase. In multiphase composites, dimensions can range from meters in materials such as steel rod-reinforced concrete structures to angstroms. In macrophase separated composite materials, properties frequently follow the rule of mixtures with the properties approximating the arithmetic mean of the properties of each individual phase, if there is good coupling between the phases. As the phases become smaller, the surface to volume ratio grows in importance with respect to properties. Interfacial and interphase phenomena begin to dominate. Surface free energies play an ever increasing role in controlling properties. In recent years, much research in materials science has been directed at multiphase systems where phase separations are submicron in at least some dimensions. It was the intent of this symposium to focus on this area of materials research.

Papers are included on blends of organic polymers, polymer-based molecular composites, and interpenetrating polymer networks. Elastomer materials such as silicone elastomers are submicron multiphase composites which have been known and used for over forty years. A review of that technology is contained in this symposium proceedings. Lately, ceramists and metallurgists have been examining similar toughening mechanisms as the polymer researchers. Hybrid materials which consist of inorganic (usually silica) and organic polymer phases with submicron phase dimensions have been reported in recent years. Papers on these new materials may also be found in this volume.

The symposium was designed to span the spectrum of synthesis, microstructure, properties, and applications of submicron multiphase materials. It was our intent, in the spirit of the Materials Research Society, to bring together researchers representing a wide variety of scientific and engineering disciplines - both established and emerging - focused on the field of submicron multiphase materials. The symposium was organized into three sections: organic organic composite materials, organic polymer/ceramic composites, and inorganic/inorganic composites including both ceramic/ceramic and metal/metal composites. Most of the papers presented at the symposium are included in this volume.

Ron Baney
Laura Gilliom
Shin-Ichi Hirano
Helmut Schmidt

June 1992

MATERIALS RESEARCH SOCIETY SYMPOSIUM PROCEEDINGS

Volume 239—Thin Films: Stresses and Mechanical Properties III, W.D. Nix, J.C. Bravman, E. Arzt, L.B. Freund, 1992, ISBN: 1-55899-133-6
Volume 240—Advanced III-V Compound Semiconductor Growth, Processing and Devices, S.J. Pearton, D.K. Sadana, J.M. Zavada, 1992, ISBN: 1-55899-134-4
Volume 241—Low Temperature (LT) GaAs and Related Materials, G.L. Witt, R. Calawa, U. Mishra, E. Weber, 1992, ISBN: 1-55899-135-2
Volume 242—Wide Band Gap Semiconductors, T.D. Moustakas, J.I. Pankove, Y. Hamakawa, 1992, ISBN: 1-55899-136-0
Volume 243—Ferroelectric Thin Films II, A.I. Kingon, E.R. Myers, B. Tuttle, 1992, ISBN: 1-55899-137-9
Volume 244—Optical Waveguide Materials, M.M. Broer, G.H. Sigel, Jr., R.Th. Kersten, H. Kawazoe, 1992, ISBN: 1-55899-138-7
Volume 245—Advanced Cementitious Systems: Mechanisms and Properties, F.P. Glasser, G.J. McCarthy, J.F. Young, T.O. Mason, P.L. Pratt, 1992, ISBN: 1-55899-139-5
Volume 246—Shape-Memory Materials and Phenomena—Fundamental Aspects and Applications, C.T. Liu, H. Kunsmann, K. Otsuka, M. Wuttig, 1992, ISBN: 1-55899-140-9
Volume 247—Electrical, Optical, and Magnetic Properties of Organic Solid State Materials, L.Y. Chiang, A.F. Garito, D.J. Sandman, 1992, ISBN: 1-55899-141-7
Volume 248—Complex Fluids, E.B. Sirota, D. Weitz, T. Witten, J. Israelachvili, 1992, ISBN: 1-55899-142-5
Volume 249—Synthesis and Processing of Ceramics: Scientific Issues, W.E. Rhine, T.M. Shaw, R.J. Gottschall, Y. Chen, 1992, ISBN: 1-55899-143-3
Volume 250—Chemical Vapor Deposition of Refractory Metals and Ceramics II, T.M. Besmann, B.M. Gallois, J.W. Warren, 1992, ISBN: 1-55899-144-1
Volume 251—Pressure Effects on Materials Processing and Design, K. Ishizaki, E. Hodge, M. Concannon, 1992, ISBN: 1-55899-145-X
Volume 252—Tissue-Inducing Biomaterials, L.G. Cima, E.S. Ron, 1992, ISBN: 1-55899-146-8
Volume 253—Applications of Multiple Scattering Theory to Materials Science, W.H. Butler, P.H. Dederichs, A. Gonis, R.L. Weaver, 1992, ISBN: 1-55899-147-6
Volume 254—Specimen Preparation for Transmission Electron Microscopy of Materials-III, R. Anderson, B. Tracy, J. Bravman, 1992, ISBN: 1-55899-148-4
Volume 255—Hierarchically Structured Materials, I.A. Aksay, E. Baer, M. Sarikaya, D.A. Tirrell, 1992, ISBN: 1-55899-149-2
Volume 256—Light Emission from Silicon, S.S. Iyer, R.T. Collins, L.T. Canham, 1992, ISBN: 1-55899-150-6
Volume 257—Scientific Basis for Nuclear Waste Management XV, C.G. Sombret, 1992, ISBN: 1-55899-151-4

MATERIALS RESEARCH SOCIETY SYMPOSIUM PROCEEDINGS

Volume 258—Amorphous Silicon Technology—1992, M.J. Thompson, Y. Hamakawa, P.G. LeComber, A. Madan, E. Schiff, 1992, ISBN: 1-55899-153-0

Volume 259—Chemical Surface Preparation, Passivation and Cleaning for Semiconductor Growth and Processing, R.J. Nemanich, C.R. Helms, M. Hirose, G.W. Rubloff, 1992, ISBN: 1-55899-154-9

Volume 260—Advanced Metallization and Processing for Semiconductor Devices and Circuits II, A. Katz, Y.I. Nissim, S.P. Murarka, J.M.E. Harper, 1992, ISBN: 1-55899-155-7

Volume 261—Photo-Induced Space Charge Effects in Semiconductors: Electro-optics, Photoconductivity, and the Photorefractive Effect, D.D. Nolte, N.M. Haegel, K.W. Goossen, 1992, ISBN: 1-55899-156-5

Volume 262—Defect Engineering in Semiconductor Growth, Processing and Device Technology, S. Ashok, J. Chevallier, K. Sumino, E. Weber, 1992, ISBN: 1-55899-157-3

Volume 263—Mechanisms of Heteroepitaxial Growth, M.F. Chisholm, B.J. Garrison, R. Hull, L.J. Schowalter, 1992, ISBN: 1-55899-158-1

Volume 264—Electronic Packaging Materials Science VI, P.S. Ho, K.A. Jackson, C-Y. Li, G.F. Lipscomb, 1992, ISBN: 1-55899-159-X

Volume 265—Materials Reliability in Microelectronics II, C.V. Thompson, J.R. Lloyd, 1992, ISBN: 1-55899-160-3

Volume 266—Materials Interactions Relevant to Recycling of Wood-Based Materials, R.M. Rowell, T.L. Laufenberg, J.K. Rowell, 1992, ISBN: 1-55899-161-1

Volume 267—Materials Issues in Art and Archaeology III, J.R. Druzik, P.B. Vandiver, G.S. Wheeler, I. Freestone, 1992, ISBN: 1-55899-162-X

Volume 268—Materials Modification by Energetic Atoms and Ions, K.S. Grabowski, S.A. Barnett, S.M. Rossnagel, K. Wasa, 1992, ISBN: 1-55899-163-8

Volume 269—Microwave Processing of Materials III, R.L. Beatty, W.H. Sutton, M.F. Iskander, 1992, ISBN: 1-55899-164-6

Volume 270—Novel Forms of Carbon, C.L. Renschler, J. Pouch, D. Cox, 1992, ISBN: 1-55899-165-4

Volume 271—Better Ceramics Through Chemistry V, M.J. Hampden-Smith, W.G. Klemperer, C.J. Brinker, 1992, ISBN: 1-55899-166-2

Volume 272—Chemical Processes in Inorganic Materials: Metal and Semiconductor Clusters and Colloids, P.D. Persans, J.S. Bradley, R.R. Chianelli, G. Schmid, 1992, ISBN: 1-55899-167-0

Volume 273—Intermetallic Matrix Composites II, D. Miracle, J. Graves, D. Anton, 1992, ISBN: 1-55899-168-9

Volume 274—Submicron Multiphase Materials, R. Baney, L. Gilliom, S.-I. Hirano, H. Schmidt, 1992, ISBN: 1-55899-169-7

Volume 275—Layered Superconductors: Fabrication, Properties and Applications, D.T. Shaw, C.C. Tsuei, T.R. Schneider, Y. Shiohara, 1992, ISBN: 1-55899-170-0

Volume 276—Materials for Smart Devices and Micro-Electro-Mechanical Systems, A.P. Jardine, G.C. Johnson, A. Crowson, M. Allen, 1992, ISBN: 1-55899-171-9

Volume 277—Macromolecular Host-Guest Complexes: Optical, Optoelectronic, and Photorefractive Properties and Applications, S.A. Jenekhe, 1992, ISBN: 1-55899-172-7

Volume 278—Computational Methods in Materials Science, J.E. Mark, M.E. Glicksman, S.P. Marsh, 1992, ISBN: 1-55899-173-5

*Prior Materials Research Society Symposium Proceedings
available by contacting Materials Research Society*

PART I

Organic/Organic Composites

FRACTURE TOUGHNESS AND FRACTURE MECHANISMS OF PBT/PC/IM BLEND

JINGSHEN WU, YIU-WING MAI AND BRIAN COTTERELL
Centre for Advanced Materials Technology
Department of Mechanical Engineering
University of Sydney
NSW 2006, Australia

ABSTRACT

Static and impact fracture toughness of a Polybutylene terephthalate (PBT)/Polycarbonate (PC)/Impact modifier (IM) blend was studied at different temperatures. The experimental results were interpreted by the specific fracture work concept and J-integral analysis. It is found that the specific fracture work concept characterizes the impact behavior of the blend very well. In the static fracture tests the specific fracture work gives the crack initiation resistance of the blend which is consistent with the J_{IC} value obtained. The effect of temperature was also examined and the fracture mechanisms were investigated via TEM and SEM. Extensive cavitation of the impact modifiers and plastic flow of matrix in the vicinity of the crack tip is believed to be the major toughening process of the enhanced fracture toughness.

INTRODUCTION

Linear Elastic Fracture Mechanics (LEFM) and J-integral analyses are now widely used to characterize fracture behaviors of polymeric materials [1]. However, LEFM requires certain restrictive size criteria of the testing specimen to be satisfied to obtain valid material constants and the J-integral method is, however, limited to static fracture analysis only. As an alternative to these methods, the specific fracture work analysis has been suggested for the determination of toughness of ductile materials [2,3]. When a crack in a ductile solid is being loaded the plastic flow occurs in an outer plastic zone bordering the fracture process zone where fracture takes place (see Fig. 1). It is necessary to separate the total fracture energy (W_f), into two parts, W_e and W_p, which are the energies dissipated in the fracture process zone and the plastic zone, respectively. Thus,

$$W_f = W_e + W_p \tag{1}$$

$$W_f = w_e B(W - a) + \beta w_p B(W - a)^2 \tag{2}$$

$$w_f = w_e + \beta w_p (W - a) \tag{3}$$

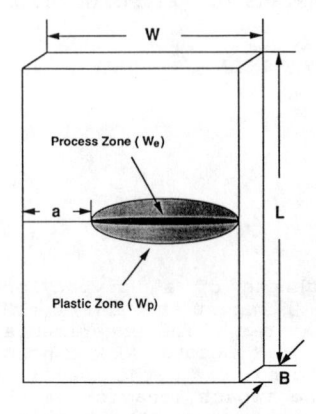

Fig. 1 Schematic of a ductile fracture specimen showing the inner fracture process zone and the outer plastic zone.

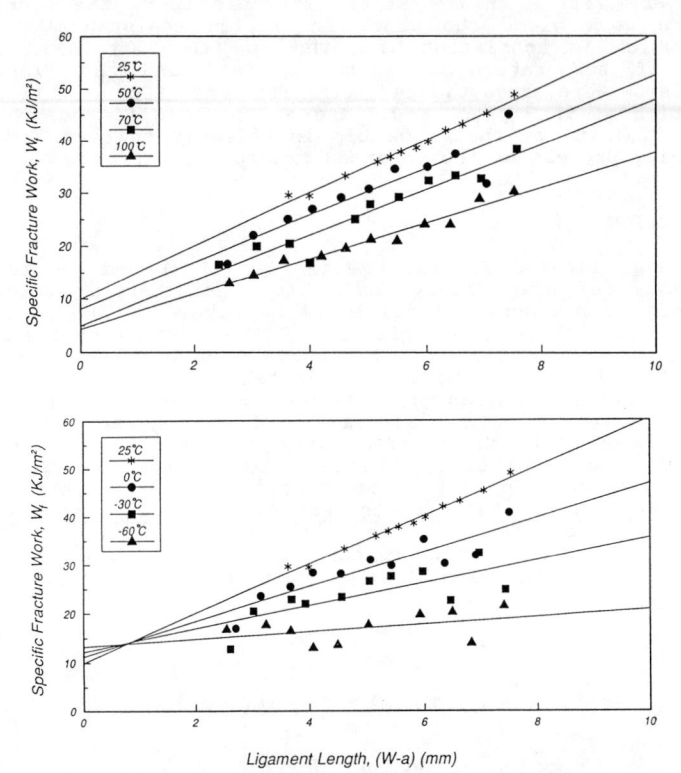

Fig. 2 Plots of specific fracture work against ligament length obtained in temperature range −60°C to 100°C.

where W_f is the specific total fracture work, w_e and w_p are the specific essential fracture work and specific plastic work, respectively; β is the plastic zone shape factor; B, W and a are thickness, width and initial crack length of specimen. Obviously, when W_f is plotted against (W - a) according to equation (3) and subsequently extrapolated to zero ligament length, w_e can be obtained from the intercept at the Y-axis and the slope of the straight line represents βw_p. It has been shown that the specific essential fracture work is a material constant for a given sheet thickness [2,3].

Substantial enhancement in toughness of a brittle polymer can be achieved by blending it with rubber particles [4]. It is suggested that the massive plastic deformation in the matrix absorbs a major part of the total fracture energy and the rubber inclusions alter the stress state in the material around the particles and induce extensive plastic deformation in the matrix. Since the role of the toughening particles is particularly important in respect to activating matrix plastic deformation the mechanical behavior of the rubber particles during fracture have received considerable attention [5-11].

In the present paper, static and impact fracture toughness of a PBT/PC/IM blend is studied at different temperatures and the toughening mechanisms involved are investigated using transmission electron microscopy (TEM) and scanning electron microscopy (SEM).

EXPERIMENTAL WORK

The material used was a commercial grade PBT/PC/IM blend supplied by Bayer AG (Australia). The specimens for Charpy impact and single-edge-notched bend (SENB) tests were cut from injection moulded plaques with the size of 6 × 8.1 × 60 mm³ and were notched on one side with a razor blade. The normalized crack length (a/W) of the specimens were varied from 0.05 to 0.75. The dimensions of the specimens for J-integral tests were 6 × 12 × 70 mm³. A deep notch with a/W = 0.5 was made in the center of the test bars. Both SENB and Charpy tests were carried out in the temperature range -196°C to 100°C. The J-integral tests were performed at 25°C, 50°C and 70°C. The SENB and J-integral tests were carried out at a rate of 5 mm/min on an Instron 4302 and the energy absorbed during displacement fracture was regarded as the area under the load-displacement curve, which was recorded digitally in a computer.

Specimens to study fracture mechanisms via TEM were prepared with two identical notches on one side of the specimen [12]. OsO_4 was used to stain the specimens which were sliced to 80~100 nm thick sections by an ultra-microtome and examined in a Philips E430. Fracture surfaces obtained from SENB tests were covered with gold before the SEM study.

RESULTS AND DISCUSSION

The results of SENB tests are shown in Fig. 2 in which the specific fracture work, W_f, is plotted against ligament length (W-a). The expected straight line relationship between W_f and (W-a) is clearly seen for the whole temperature range

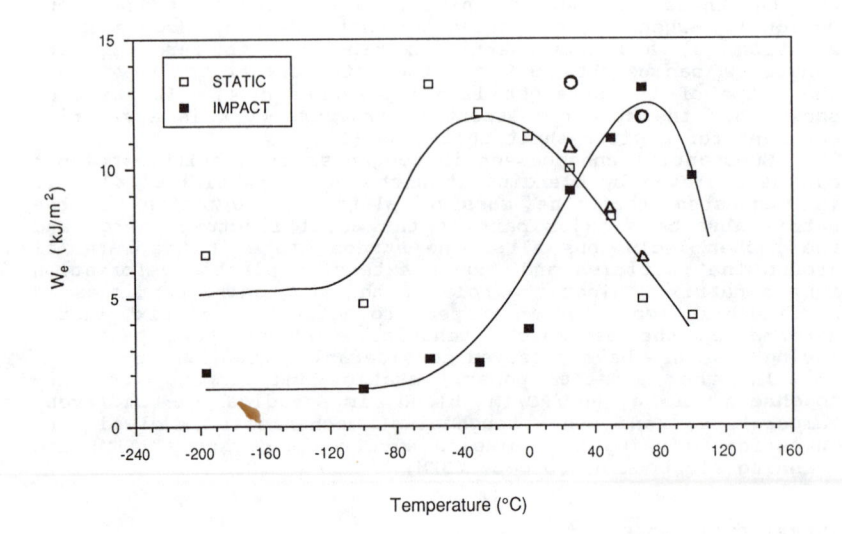

Fig. 3   Variation of specific essential fracture work with temperature for both static and impact tests (△ J_{IC-81}; ○ J_{IC-89}).

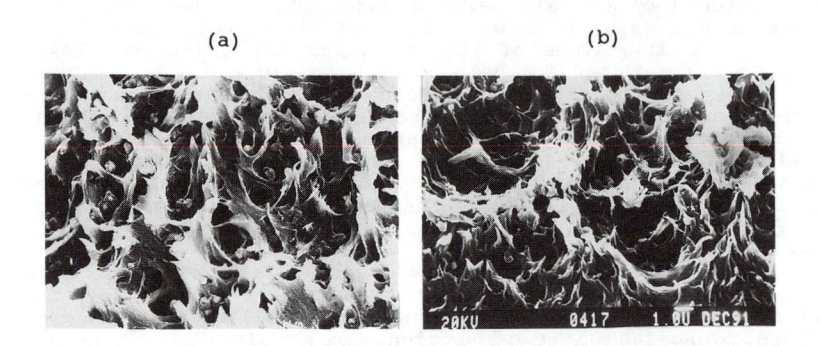

Fig. 4   SEM photograph taken from a fracture surface obtained in static SENB test at -196°C (a) shows the same features as the fracture surface obtained in impact test at -30°C (b).

and the scatter of the data points is small. The specific essential fracture work, w_e, can be obtained easily by extrapolating the straight line relationship to zero ligament length. The specific plastic work, βw_p, can also be easily determined from the slope of the lines. It is also clear that in the temperature range 25°C to -60°C the slope of the lines decreases gradually with decreasing temperature, implying that βw_p is smaller at a lower temperature. Since w_p must increase with decreasing temperature this means that β or the plastic zone size must decrease with temperature. In fact it is expected that at very low temperature the plastic flow in the ligament is negligible (i.e. $\beta \rightarrow 0$) and a horizontal line is obtained when w_f is plotted against (W-a). This means that $w_f = w_e$ under this condition. In the temperature range 25°C to 100°C, however, βw_p decreases with increasing temperature because w_p decreases with temperature even though β remains unchanged in this temperature range.

The critical J-integral, J_{IC}, of the blend was measured at three test temperatures using two ASTM standard methods, E813-81 (J_{IC-81}) and -89 (J_{IC-89}). The impact fracture energies recorded during Charpy tests at different temperatures were also interpreted by means of the specific fracture work analysis since the J-integral method is limited to static fractures only.

The variations of impact and static fracture toughness, w_e, with temperature is plotted in Fig. 3 together with the values of J_{IC-81} and J_{IC-89}. It is noted that: (1) The values of J_{IC-81} are much smaller than those of J_{IC-89}. This is because J_{IC-89} is the J value after a crack growth of 0.2 mm, whereas J_{IC-81} is the value at crack initiation. (2) The values of J_{IC-81} agree extremely well with those of w_e, suggesting that both J_{IC-81} and w_e give the crack initiation toughness. These experimental results also verified the equivalence of the two methods used. (3) The profile of the curve of static fracture toughness with temperature are similar to the variation of impact fracture toughness with temperature, except that the impact curve is shifted to a higher temperature range. This means that the activation of the toughening processes is postponed to a higher temperature range because of the high strain rate in impact tests and time-temperature superposition holds for the fracture toughness-temperature relationship.

SEM study of the fracture surfaces reveals that the low temperature statically fractured surfaces (Fig. 4a) possess the same features as the impact fractured surfaces obtained at high temperature (Fig. 4b), indicating that both static and impact fracture toughness enhancement originates from the same source, which is proven to be the plastic blunting of crack tip caused by the relaxation of rubbery particles at T_g and the β transition loss process of the parent polymer.

TEM study substantiates the same conclusion mentioned above. The deformation behavior of the samples in static test at 25°C (Fig. 5) are similar to those impact deformed at 70°C (Fig. 6). Moreover, the sequence of the toughening events have been clearly demonstrated in these pictures. That is, first cavitation and/or crazing without shear yielding, then dilatation of the cavitated particles together with some extensive shear in the matrix, then massive shear flow in the matrix and finally, very large elongation of the rubbery inclusions together with severe crack tip blunting.

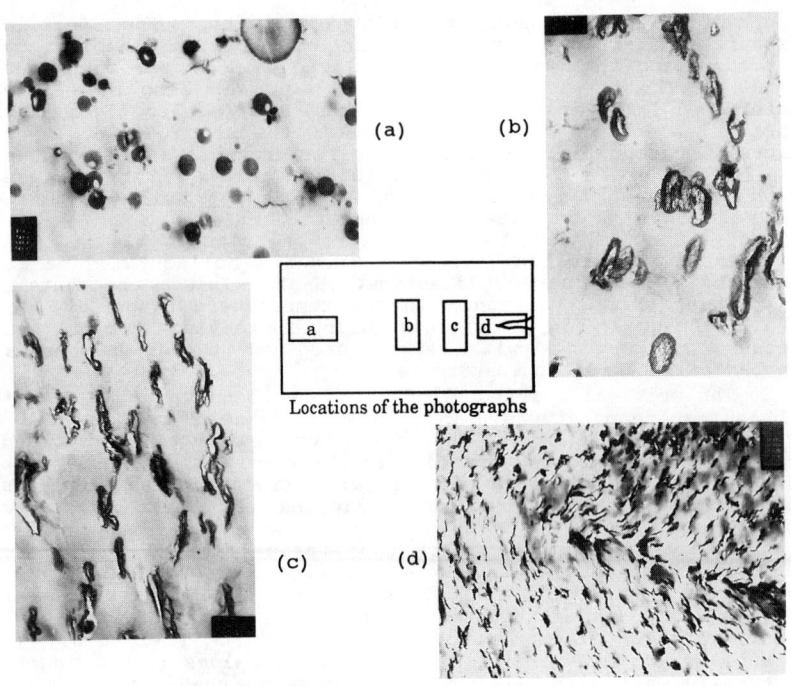

Fig. 5 TEM micrographs taken from an ultra-thin section of a statically loaded sample at room temperature.

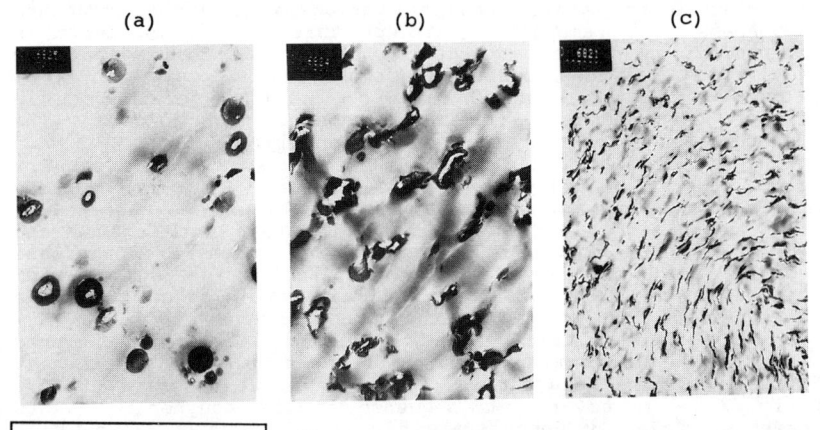

Fig. 6 TEM micrographs taken from an ultra-thin section of an impact sample at 70°C. Fracture mechanisms are seen to be cavitation and shear yielding. No crazing is found.

ACKNOWLEDGEMENT

The authors wish to thank Bayer AG (Australia) for the supply of the PBT/PC/IM resins. One of us (JSW) is supported by a Sydney University Postgraduate Research Award. R.P. Burford kindly provided the use of an injection moulder for manufacture of specimens for testing.

REFERENCES

1. J.G. Williams, Fracture Mechanics of Polymers (Ellis Horwood Limited, Chichester, 1984).

2. Y-W Mai and B. Cotterell, Int. J. Fract. 32, 105 (1986).

3. Y-W Mai, Essential work of fracture and J-integral measurements of ductile polymers, in Proc. of Inter. Symp. on How to Improve the Toughness of Polymers and Composites (Yamagata University, Yamagata, Japan, October 8, 1990).

4. C.B. Bucknall, Toughened Plastics, (Applied Science Publishers Ltd., London, 1977).

5. A.M. Donald and E.J. Kramer, J. Appl. Polymer Sci., 27, 3729 (1982).

6. S. Wu, Polymer, 26, 1855 (1985).

7. S. Wu, J. Polym. Sci., Polym. Phys. Edn., 21, 699 (1983).

8. R.J.M. Borggreve, R.J. Gaymans, J. Schuijer and J.F. Ingen-Housz, Polymer, 28, 1489 (1987).

9. A.F. Yee and R.A. Pearson, J. Mater. Sci., 21, 2462 (1986).

10. R.A. Pearson and A.F. Yee, J. Mater. Sci., 21, 2475 (1986).

11. D.S. Parker, H-J Sue, J. Huang and A.F. Yee, Polymer, 30, 570 (1989).

12. H-J Sue and A.F. Yee, J. Mater. Sci., 24, 1447 (1989).

A NOVEL EPOXY RESIN TOUGHENED BY HTBN LIQUID RUBBER

XIAOZU HAN, ZHANKUI YUN AND FENCHUN GUO
Changchun Institute of Applied Chemistry, Academia Sinica, Changchun 130022, China

ABSTRACT

A novel toughened epoxy resin was obtained by using an epoxy-terminated prepolymer prepared from epoxy resin and hydroxy-terminated butadiene-acrylonitrile copolymer (HTBN), and an amine curing agent. The cured, toughened resin has excellent mechanical properties due to the two-phase structure which was observed using SEM and TEM. When the HTBN content is 15 phr, the rubber phase separates effectively and the specimen presents a fine two-phase structure. However, when the HTBN content reaches 25 phr, the morphology appears continuous.

INTRODUCTION

The mechanical properties of epoxy resin can be significantly improved by adding a reactive liquid rubber. Carboxy-terminated butadiene-acrylonitrile copolymer (CTBN) is conventionally used (1,2). The price of CTBN is, however, higher than that of other liquid rubbers. Hydroxy-terminated butadiene-acrylonitrile copolymer (HTBN), a cheaper liquid rubber, was used for modifying the epoxy resin with m-phenylenediamine as curing agent, but the toughening efficiency was lower than that of CTBN (3). In this case, the hydroxyl groups in HTBN do not readily react with epoxy groups, so the rubber chains cannot form a cohesive network with the epoxy resin.

In order to bring the HTBN chains into the epoxy resin network, an epoxy resin/HTBN/diacid anhydride system was investigated (4). The diacid anhydride can react with both the hydroxy group on HTBN and the epoxy group on the epoxy resin to form one network, thus significantly improving the properties of the epoxy resin. In another method, isocyanate-terminated prepolymer (ITBN) obtained by the reaction of HTBN with an excess of tolulene diisocyanate (TDI) was used as the toughening agent, with primary diamine used as the curing agent (5). The amino group can react with both the isocyanate group on ITBN and the epoxy group to bond the rubber chains to the epoxy resin network. The toughness of the epoxy resin could be increased, but the pot life of the system was a bit too short for practical application.

In this paper, ITBN was obtained at first from the reaction of HTBN and TDI and then reacted with hydroxyl groups on excess epoxy resin to form an epoxy-terminated prepolymer (ETBN). The ETBN can be cured by conventional amine curing agents. The toughened epoxy resin is convenient to use and has excellent mechanical properties.

EXPERIMENT

Materials

Bisphenol A epoxy resin (E-51) was supplied by Shanghai Resin Factory. The epoxy group content is 0.52 mol/100g and the hydroxyl group content is 0.040 mol/100g.

HTBN with 15% acrylonitrile was prepared by free radical polymerization using hydrogen peroxide as the initiator at 117° C. The molecular weight is about 3000 and the hydroxyl group content 0.07 mol/100g.

TDI and the curing agents, including m-xylene diamine (MXDA), 4,4'-diamino-diphenyl methane (DADM), m-phenyldiamine (MPDA), were obtained from standard sources.

Syntheses of ITBN and ETBN

One equivalent HTBN was reacted with two equivalents TDI at 80° C to obtain ITBN. A certain amount of ITBN was added to the E-51 epoxy resin and the mixture was heated at 85°C for 1.5h to obtain ETBN. The reactions that occured are schematically shown below:

The isocyanate groups on ITBN reacted with the hydroxyl groups on the epoxy resin in the formation of ETBN. Because the hydroxyl group content is very low, most of the epoxy resin molecules have no hydroxyl group, so many unreacted epoxy resin molecules in ETBN remain.

The reaction betweeen ITBN and the epoxy resin went to completion, which was proven by the IR spectra. The reaction temperature is very important in determining the structure of ETBN obtained. When the temperature is higher than 140° C, the isocyanate group can react with an epoxy group to form an oxazolidone ring (6).

Preparation of toughened epoxy resin

100 parts of ETBN were mixed with a certain amount of curing agent (e.g. 20 parts). The mixture was vacuum-degassed and then poured into a mold and cured in an oven at different temperatures to obtain cured specimens for testing.

Measurement

The mechanical properties, including stress-strain behavior and lap shear strength, were measured by an INSTRON 1121. The fracture toughness of the sample was calculated from the area under the stress-strain curve. The glass transition temperature (T_g) was determined using differential scanning calormetry.

Morphologies of cured specimens were examined with scanning electron microscopy (SEM, Model JXA-840) and transmission electron microscopy (TEM, Model H-500). The SEM specimens were obtained from the fracture surfaces, coated with silver. The TEM specimens were cut by ultramicrotomy and stained with OsO_4.

RESULTS AND DISCUSSION

Effects of HTBN content on the properties of cured resin

When MXDA and MPDA-DADM were used as curing agents with the level being 20 phr, the effects of HTBN content on the properties of cured resin are shown in Fig. 1 and Fig. 2. It is clear that the tensile strength decreased with increasing HTBN content; however, the shear strength, elongation and toughness of cured resin appeared to be maximum when the HTBN contents were 15-25 phr, while the toughening efficiencies decreased when the HTBN content exceeded 25 phr.

The T_gs of toughened epoxy resin changed to a certain extent with a change in the HTBN content (see Table I). When HTBN contents were 0-15 phr, T_gs did not change appreciably, but when HTBN content exceeded 20 phr, the T_gs decreased.

Table II lists the shear strengths and peel strengths of epoxy resins toughened by HTBN and CTBN, compared to one containing no toughening agent. It illustrates that HTBN has a greater toughening efficiency than that of CTBN.

Table I: Tgs of toughened resins

HTBN level, phr	0	7.5	15	20	25
Tg, °C	81.7	80.9	83.8	71.4	72.7

Table II: Effects of toughening method on strengths

Strength	Untoughened	Toughened by 25 phr CTBN	Toughened by 25 phr HTBN
Shear strength, MPa	19.4	23.8	35.5
Peel strength, N/cm	2.0	20.6	41.2

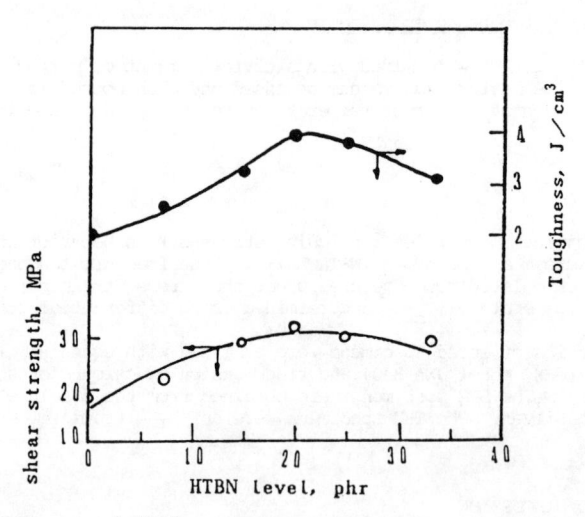

Fig.1 Effects of HTBN content on properties of cured resin
(MXDA as curing agent; 120℃, 16 h, 125℃, 4 h)

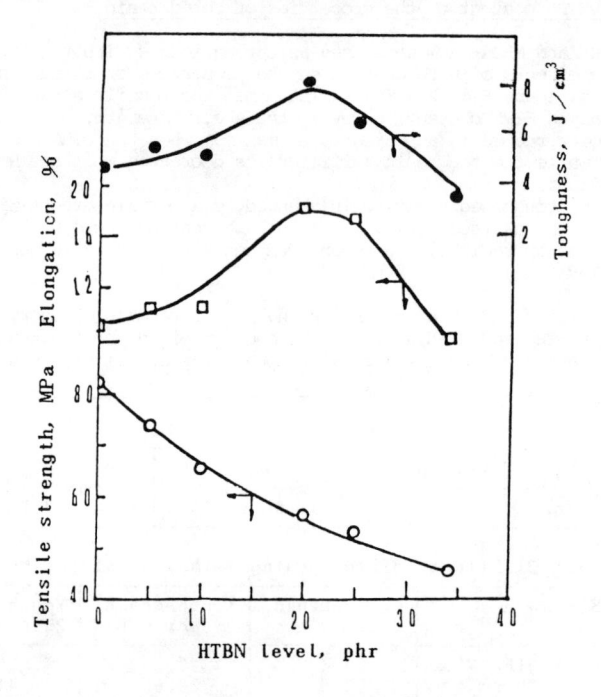

Fig.2 Effects of HTBN content on properties of cured resin
(MPDA—DADM as curing agent; 120℃, 16 h, 125℃, 4 h)

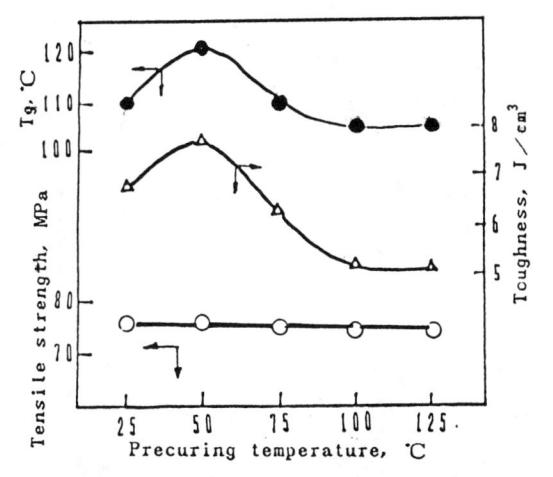

Fig.3 Effects of precuring temperature

Effects of precuring temperature

The mixture of ETBN and MPDA-DADM curing agent was pre-cured at different temperatures for 16h and then cured at 125° C for 4h. The effects of precuring temperature on the properties of the cured resin is shown in Fig. 3. It is observed that the precuring temperature does not influence the tensile strength, but it significantly affects the toughness and T_g. When the precuring temperature is about 50°C, both the toughness and T_g are increased. The more obvious toughening efficiency achieved at this temperature may be attributable to the fact that micro-phase separation occurs more favorably at lower temperatures.

Morphology of toughened resin

Phase separation is of importance in epoxy resin systems toughened by liquid rubbers (1). We have investigated the morphologies of HTBN-toughened epoxy resin using SEM and TEM. The results are shown in Fig. 4.

When the HTBN contents are 7.5 and 15 phr, the rubber phase separates effectively with domain sizes of 0.1 to 2 μm, which are smaller than those of toughened epoxy resins studied previously (4). An interesting point is that the dispersed phase still presents a fine two-phase structure, which means that some epoxy resin chains are inserted into the rubber phase. When the HTBN contents reach 25 and 35 phr, the morphologies appear continuous and the mechanical properties of the resin are degraded.

REFERENCES

(1) C.K. Riew, E.H. Rowe, A.R. Siebert, Adv. Chem. Ser., 154, 326 (1976).
(2) A.K. Banthia, et al., ibid., 222, 343 (1989).
(3) D.A. Scola, U.S., 3, 926, 904 (1975).
(4) X. Han, S. Li, Q. Zhang, Chinese J. Polm. Sci., 8, 335 (1990).
(5) S. Sankaran, M. Chanda, J. Appl. Polym. Sci., 39, 1459 (1990).

SEM　　　　　　　　　　TEM

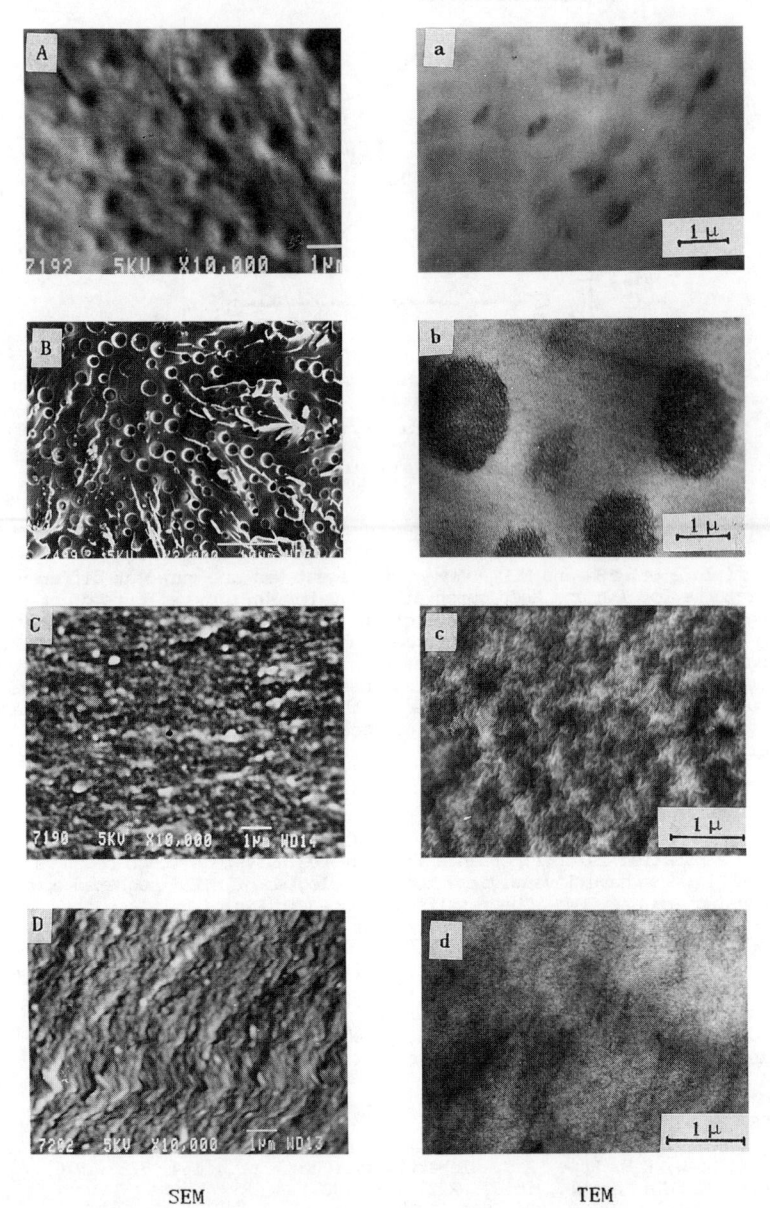

Fig.4 Morphologies of HTBN toughened epoxy resin
HTBN level, phr: A, a, 7.5; B, b, 15;
C, c, 25; D, d, 35

STRUCTURE-PROPERTY RELATIONSHIPS IN THE TOUGHENING OF POLY(METHYL METHACRYLATE) BY SUB-MICRON SIZE, MULTIPLE-LAYER PARTICLES

A.C. ARCHER, P.A. LOVELL*, J. McDONALD, M.N. SHERRATT & R.J. YOUNG
Polymer Science and Technology Group, Manchester Materials Science Centre, UMIST, Grosvenor St., Manchester, M1 7HS, United Kingdom

ABSTRACT

Rubber-toughened poly(methyl methacrylate) materials have been prepared by blending poly(methyl methacrylate) with specially-synthesised, refractive index matched particles containing two, three and four radially-alternating rubbery and glassy layers. The paper describes the effects upon mechanical properties of (i) two-, three- and four-layer particle structure and (ii) particle diameter and glassy core size for three-layer particles.

INTRODUCTION

Rubber-toughened poly(methyl methacrylate) (RTPMMA) materials are produced by blending PMMA with separately-prepared toughening particles [1]. Emulsion polymerisation is used to prepare the toughening particles which typically comprise two to four radially-alternating rubbery and glassy layers, the outer layer always being of glassy polymer. The particles are crosslinked during their formation in order to ensure that they retain their morphology and size during blending with PMMA and during subsequent moulding of the blends. This route to RTPMMA has the distinct advantage of allowing independent control of the properties of the matrix PMMA, the composition, morphology and size of the dispersed rubbery phase, and the level of inclusion of the toughening particles. This paper presents some results from a major research programme aimed at elucidating the mechanism(s) of deformation, and optimising mechanical properties, of RTPMMA materials.

EXPERIMENTAL

Materials

The matrix PMMA used in preparing the blends was Diakon LG156 (ICI plc) which is poly[(methyl methacrylate)-co-(n-butyl acrylate)] with 8 mol% n-butyl acrylate repeat units and has M_n = 47 kg mol^{-1} and M_w/M_n = 1.7 (GPC, polystyrene calibration).

The two-, three- and four-layer (i.e. 2L, 3L and 4L) toughening particles are represented schematically in *Figure 1* and were prepared by sequential emulsion polymerisation [2]. In order to retain a high percentage transmission of visible light in the RTPMMA materials, the compositions of the rubbery and glassy phases of the particles were chosen such that their refractive indices match that of the matrix PMMA. The rubbery layers consist of crosslinked poly[(n-butyl acrylate)-co-styrene] and the glassy layers consist of poly[(methyl methacrylate)-co-(ethyl acrylate)], the inner glassy layers being crosslinked. Each layer is graftlinked to the other layers with which it is in contact. With the exception of the 3LAI and 3LAII particles, allyl methacrylate (ALMA) was used at constant levels for both crosslinking and graftlinking. In the preparation of the 3LAI particles, ALMA was used at higher-than-normal levels in the interfacial regions in order to increase the interfacial strength, whereas the 3LAII particles were prepared with ALMA at normal levels in the interfacial regions but replacing ALMA for crosslinking the bulk of the glassy and rubbery phases by identical molar quantities of ethan-1,2-diol dimethacrylate and hexan-1,6-diol diacrylate respectively.

For each type of toughening particle, the latex obtained from emulsion polymerisation was coagulated by addition to magnesium sulphate solution to yield loose aggregates of the particles. These were isolated by filtration, washed thoroughly with water and then dried at 70 °C. The

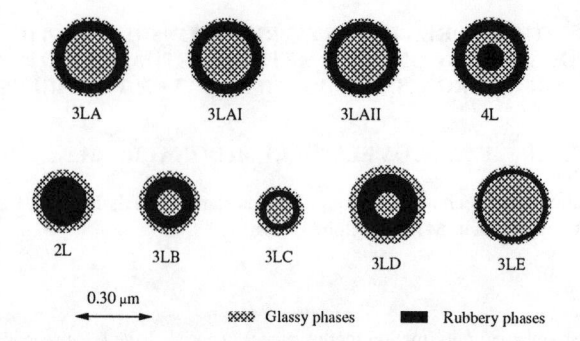

Figure 1. Schematic diagrams of sections through the equators of the toughening particles showing their sizes and internal structures.

dried aggregates of toughening particles were blended with the matrix PMMA at 220 °C by either a single pass through a Werner Pfleiderer 30 mm twin-screw extruder or two passes through a Francis Shaw 40 mm single-screw extruder. This ensured complete disruption of the aggregates to give uniform dispersions of the toughening particles, as evidenced by transmission electron micrographs of ultramicrotomed sections of the blends. For each type of particle, four blends containing different weight fractions (w_p) of particles were produced. The nomenclature used to define the blends is illustrated by 3LA22 which specifies that the blend contains 3LA particles with $w_p = 0.22$.

Testing

The matrix PMMA and each of the RTPMMA materials were compression moulded into plaques of 3 mm thickness (for tensile testing) and of 6 mm and 12 mm thickness (for fracture testing). Tensile stress-strain data were obtained at 20 °C according to ASTM D638-84 employing a nominal strain rate of 2×10^{-3} s^{-1}. Plane strain values of critical strain energy release rate (G_{Ic}) were determined in Charpy mode with an impact velocity of 0.85 m s^{-1} using a Ceast instrumented-pendulum impact tester and employing procedures defined in a recently published protocol for fracture testing of polymers [3]. In order to observe the mode of crack propagation, single and double edge-notched specimens were subjected to sub-critical crack growth in 3-point bend mode and then either (i) polished from both faces to obtain thin sections for examination in an optical microscope, or (ii) polished from one face followed by ultramicrotoming and then staining with ruthenium tetroxide to obtain sections for examination by transmission electron microscopy (TEM). Fracture surfaces were also examined by scanning electron microscopy (SEM).

RESULTS AND DISCUSSION

Tensile testing revealed that, with the exception of 3LAII particles, each type of toughening particle is capable of inducing large-scale yielding in the matrix PMMA leading to much greater fracture strains (up to 30-50%, compared to 3% for the unmodified matrix PMMA) and greatly increased energies to fracture. The poor performance of the 3LAII materials (ca. 8% ultimate elongation for $w_p > 0.20$) results from the change in the chemistry of crosslinking in 3LAII particles. Although the other types of particle give ductile RTPMMA materials, there are significant differences in deformation behaviour. These are discussed below in terms of the results from tensile testing, fracture testing and microscopic examination of deformation zones.

The values of Young's modulus (E) and yield stress (σ_y) obtained from tensile testing are shown in *Figure 2*, and will be considered together with the impact values of critical strain energy release rate (G_{Ic}) shown in *Figure 3*. In each case, the data are plotted against (i) particle volume fraction (V_p), which excludes the outer glassy layer of particles but, for the three- and four-layer particles, includes internal glassy phases, and (ii) volume fraction of rubber (V_r)

which excludes both the outer and any internal glassy phases. The values of G_{Ic} published previously for the 2L, 3LA and 4L materials [2a,4] are higher than those reported here. The earlier values were determined using an instrumented, falling dart instrument with an impact velocity of 1 m s^{-1}; pre-cracks were sharpened just prior to testing by tapping a fresh razor blade into the pre-formed notch. In the present work, the latter procedure was found to give significantly higher values of G_{Ic} than when sharpening the notch using a sawing action, which was the procedure used for the measurements reported in this paper. Similar findings have been reported in studies of other materials [3].

Inspection of *Figure 2* shows that E and σ_y are largely controlled by V_r with some small affects arising from particle size/morphology. In contrast, examination of *Figure 3* reveals that G_{Ic} depends strongly upon particle size/morphology.

Comparison of the results for the 2L and 3LB materials reveals that the effect of introducing a glassy core into a homogeneous rubbery particle is to increase E, σ_y and G_{Ic} for a given level of particles (i.e. V_p). Thus, in the 3LB materials, the glassy core makes the rubber much more effective in improving toughness, as can clearly be seen in *Figure 3b*. Although the value of E appears to depend only on V_r, there is evidence that σ_y is slightly greater for 3LB materials than for 2L materials of similar V_r.

The effects of particle size are apparent from comparison of the properties of the RTPMMA materials prepared from the 3LB, 3LC, and 3LD particles, which have 100 nm diameter glassy cores with rubbery-layer diameters of 202, 152 and 255 nm respectively. The values of E and σ_y show no dependence upon particle size. The higher values of E for the 3LC materials at given values of V_p, arise from the higher percentage of glassy phase within the 3LC particles, as is evident from *Figure 2b*. There is, however, a strong dependence of G_{Ic} upon particle size, with the smallest particles (3LC) giving rise only to small increases in G_{Ic} compared to the matrix PMMA, even at high V_p. The increase in size from 3LC to 3LB particles leads to a major increase in G_{Ic}. The further increase in size from 3LB to 3LD particles leads to a much smaller increase in G_{Ic} at a given value of V_p, the increase arising mainly from the increased volume fraction of rubber but also from a particle size effect since the increase in G_{Ic} is disproportionately larger at higher V_p (compare *Figures 3a and 3b*).

By considering the mechanical properties of the 3LA, 3LD and 3LE materials it is possible to examine the effects of glassy core diameter (200, 100 and 244 nm respectively) in three-layer toughening particles with approximately constant rubbery-layer diameter (284, 255 and 290 nm respectively). The increase in glassy core size leads to major increases in E, σ_y and G_{Ic} for a given value of V_p, though the increases in E and σ_y are due essentially to decreases in V_r. The increase in G_{Ic}, however, is more evident when considered in terms of V_r, showing that increasing the glassy core size leads to more efficient use of the rubber. Hence, for the 3LE materials, the increase in G_{Ic} is achieved with relatively small reductions in E.

The introduction of a rubbery core into the inner glassy phase of 3LA particles has no major effect upon mechanical properties, and would appear to have advantage only in terms of circumventing the earlier patents covering these types of RTPMMA materials (compare *Ref. 1c* with *Refs. 1a* and *1b*).

In addition to the effects of particle size/morphology, the effects of changing the crosslinking and graftlinking in the 3LA particles have been examined. The higher level of graftlinking at interfaces in the 3LAI particles leads to modest increases in E, σ_y and G_{Ic}. The change of crosslinking agent from ALMA to hexan-1,6-diol diacrylate for the rubbery phases in the 3LAII particles, however, causes a dramatic loss of ductility with major decreases in G_{Ic} to values not much greater than that of the matrix PMMA, though there is no major effect upon E.

Preliminary studies of the mechanism(s) of deformation operating in the RTPMMA materials have been carried out, but at present are not sufficiently complete to be definitive. Nevertheless, some of the more important observations from these studies are discussed below.

Significant differences in deformation behaviour are evident in tensile tests. The 2L, 3LA, 3LAI, 3LE and 4L materials develop diffuse shear bands and stress-whiten at, and beyond, the yield point. The 3LB and 3LD materials give coarse shear bands emanating from diamond-shaped features which form just prior to yield; additionally, they develop an undulating surface texture and stress-whiten. The 3LC materials also show coarse shear bands emanating from diamond-shaped features and have surface texture, but do not significantly stress-whiten. Examples of the different types of shear banding are shown in *Figure 4*.

SEM of fracture surfaces from tensile and single edge-notched, 3-point bend specimens of 2L, 3LA and 4L materials shows the presence of numerous holes and dome-like features with diameters similar to, and also smaller than, those of the toughening particles, suggesting that

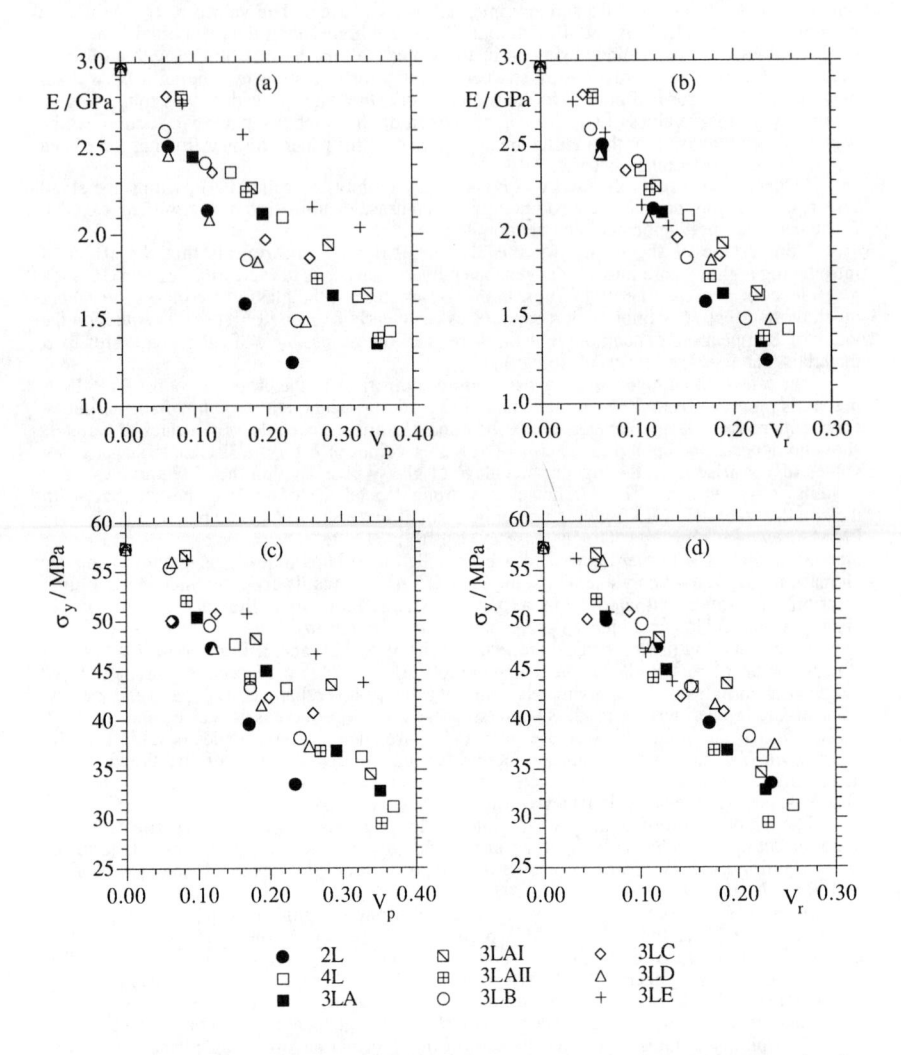

Figure 2. Plots of Young's modulus (E) and yield stress (σ_y) against particle volume fraction (V_p) and volume fraction of rubber (V_r) for each of the RTPMMA materials. [Fracture stresses have been plotted in (c) and (d) for the matrix PMMA, the 3LAII materials, 2L11 and 3L10 because these materials did not yield.]

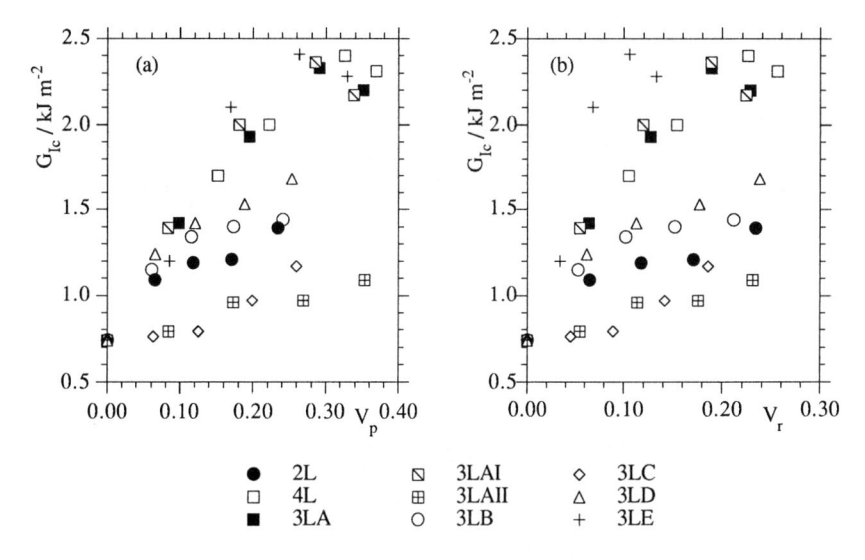

Figure 3. Variation of 0.85 m s^{-1} impact values of the critical strain energy release rate (G_{Ic}) with (a) particle volume fraction (V_p) and (b) volume fraction of rubber (V_r) for each of the types of RTPMMA.

cavitation and/or debonding of the toughening particles contributes to stress-whitening. However, TEM of sections taken from just below fracture surfaces of tensile and single edge-notched, 3-point bend specimens of 3LA30 and 4L25 indicate that cavitation is not extensive.

Optical microscopy of deformation zones formed at the tips of sub-critically grown cracks in the RTPMMA materials shows flame-shaped plastic zones. The cores of the plastic zones are highly birefringent in samples obtained at low rates of deformation and are surrounded by fine dark bands, as shown in *Figure 5a*. TEM of the birefringent region within the plastic zone shows little evidence of particle cavitation but reveals that the particles are highly distorted, the degree of distortion decreasing towards the edge of the zone (*Figures 5b* and *5c*). These features are very similar to those reported for nylon 6.6/poly(phenylene oxide) blends by Sue and Yee [5], who concluded that initially a crazed zone forms ahead of the crack tip and then transforms into a shear yielded zone as the crack propagates through it, the crazes in the original crazed zone being closed or distorted by the shear yielding process.

The TEM micrographs shown in *Figure 6* are of the deformation around and ahead of a crack grown sub-critically under impact conditions in 3LE30. *Figure 6a* is a micrograph taken at relatively low magnification in the plane normal to the crack and shows that the particles are connected, through the matrix, by fine dark bands which are non-birefringent when viewed in a polarising optical microscope. At higher magnification (*Figure 6b*), the bands can be seen to have propagated between the particle equators, with no evidence of particle cavitation. Since the bands are non-birefringent and are associated with positions of maximum triaxial stress concentration, they may be considered to be microcracks or, possibly, microcrazes. A further feature in the deformation of 3LE30 is the formation of microvoids in the plane of the crack, ahead of the plastic zone (*Figure 6c*).

CONCLUSIONS

The results presented in this paper show that the toughness of RTPMMA under impact is strongly dependent upon the size and morphology of the toughening particles, but that Young's

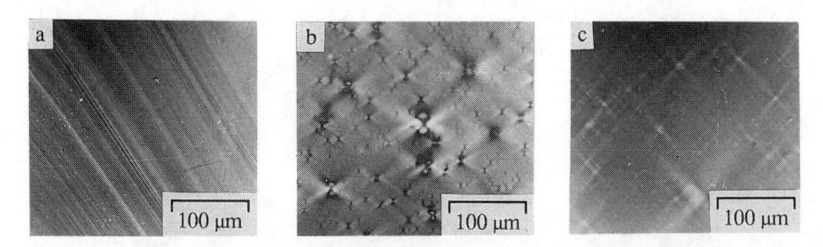

Figure 4. Optical micrographs from fractured tensile specimens showing shear deformation bands viewed between crossed, polarising filters to enhance contrast between the birefringent bands and the undeformed matrix: (a) bands typical of those observed in 2L, 3LA, 3LAI, 3LE and 4L materials, (b) bands typical of those observed in 3LB and 3LD materials, and (c) bands observed in 3LC materials.

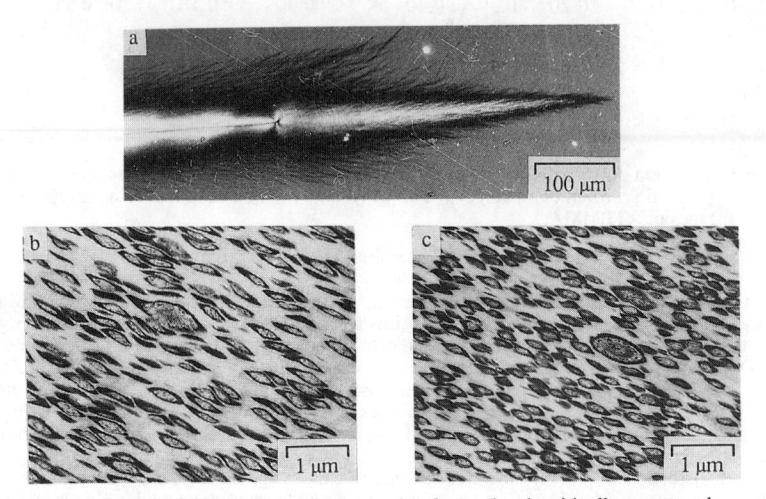

Figure 5. Micrographs of the deformation at the tip of a crack sub-critically grown at low rate in a single edge-notched, 3-point bend specimen of 3LA30. (a) Optical micrograph of the plastic zone taken using polarising filters crossed at 45°. (b) TEM micrograph of deformed particles within the birefringent region of the plastic zone. (c) TEM micrograph of less severely deformed particles at the edge of the zone.

modulus and yield stress are largely controlled by the volume fraction of rubber. Toughness increases as rubbery-layer diameter increases from approximately 150 to 200 nm, though for diameters above 200 nm the effect of size is much less significant. The additional constraints arising from the presence of a glassy core within a rubbery particle give rise to significant increases in toughness, the increase becoming greater as the size of the glassy core is increased for the materials studied. Of the particles investigated, the optimum are the 3LE type, which have a large glassy core and a rubbery-layer diameter of 290 nm. These particles give RTPMMA materials which have the highest Young's modulus and yield stress for a given value of critical strain energy release rate.

The chemistry of crosslinking of the rubbery phases in the particles is important. When allyl methacrylate is replaced by an equimolar quantity of a hexan-1,6-diol diacrylate, a dramatic decrease in toughness is observed, most probably as a consequence of the greater efficiency (i.e. higher degree) of crosslinking [6].

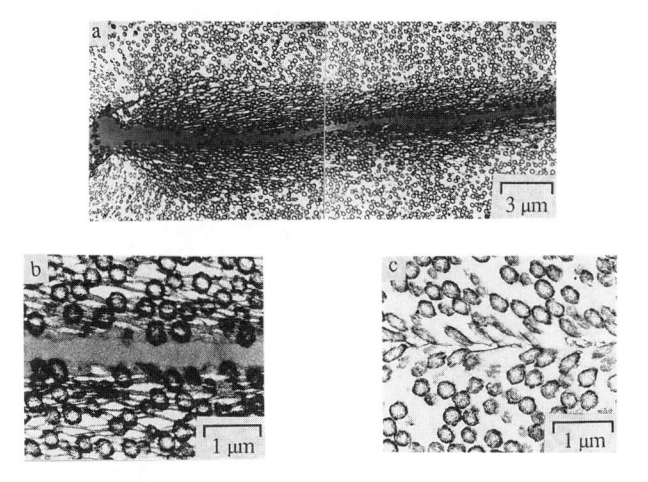

Figure 6. TEM micrographs of the deformation around and ahead of a crack sub-critically grown under impact in a double edge-notched, 3-point bend specimen of 3LE30. (a) Low magnification micrograph showing fine deformation bands in the plane normal to the crack. (b) The bands in (a) shown at higher magnification. (c) Microvoids formed ahead of and in the plane of the crack shown in (a).

The micromechanics of deformation in the RTPMMA materials have not yet been studied sufficiently to provide a definitive picture of the mechanism(s) of deformation operating in the materials. Furthermore, the differences in shear bands and stress-whitening in tensile specimens between the different types of RTPMMA materials suggest that the deformation mechanism(s) may change with particle size and morphology. Nevertheless, there is strong evidence that the particles induce voiding prior to undergoing shear deformation, and that this voiding arises, at least in part, from particle cavitation and/or debonding, with the possibility of significant contributions from microcracks generated between particles. Further microscopy work is necessary, however, before firm conclusions can be made about the mechanism(s) of deformation in the RTPMMA materials.

ACKNOWLEDGEMENTS

The authors express their thanks to the Science and Engineering Research Council and ICI plc for funding this research programme. The assistance of Pat Hunt, Mike Chisholm and Bill Jung of ICI plc is gratefully acknowledged.

REFERENCES

1. (a) Rohm and Haas Company, Brit. Patent No. 1 340 025 (1973); (b) Rohm and Haas Company, Brit. Patent No. 1 414 187 (1975); (c) E.I. Du Pont de Nemours and Company, Gt. Brit. Patent Appl. No.2 039 496A (1979)
2. (a) P.A. Lovell, J. McDonald, D.E.J. Saunders, M.N. Sherratt and R.J. Young, Advances in Chemistry Series 233, American Chemical Society, Washington, in press; (b) P.A. Lovell, J. McDonald, D.E.J. Saunders and R.J. Young, submitted to Polymer
3. J.G. Williams, EGF Fracture Testing Protocol, March 1990
4. P.A. Lovell , J. McDonald, D.E.J. Saunders, M.N. Sherratt and R.J. Young, Plast. Rubb. Comp. Proc. Appl. 16, 37 (1991)
5. H-J. Sue and A.F. Yee, J. Mat. Sci., 24, 1447 (1989)
6. F. Heatley, P.A. Lovell and J. McDonald, submitted to Eur. Polym. J.

A STUDY ON THE BLENDS OF EPOXY/POLYBUTADIENE AND THE APPLICATION TO THE ENCAPSULATION OF A CAPACITOR

ZHONGYUAN REN AND LIYING QUI
Xi'an Jiaotong University, Chemical Engineering Dept., People's Republic of China

ABSTRACT

This paper describes the blends of epoxy/polybutadiene and the application of the blends to the encapsulation of capacitors. Experiments showed that the hydroxy-carboxyl terminated polybutadiene (HCTPB) had a good toughening effect on epoxy resins, and the blends of epoxy/HTPB or epoxy/HCTPB had good craze resistance at low temperatures. The phase separation and dynamic mechanical analysis of these blends are discussed below.

INTRODUCTION

Epoxy resin is a superior thermosetting material. It is widely used in mechanical, electrical, electronic, aeronautical and space technologies. Unfortunately, the cured epoxy resin itself is brittle and has poor craze resistance, which limits the use of epoxy resin. The toughening modification of epoxy resin is therefore an important research subject.

In the 1960's, McGarry used CTBN to successfully toughen epoxy resin. Since then, many research reports have been published on the modification of epoxy resin by polyether, polyester, polyurethane resins and silicone rubber. In our research work, the toughening effect of thiokol and its application to the encapsulation of capacitors has also been studied.

This paper discusses epoxy resin blends toughened with two kinds of polybutadiene: hydroxyl-terminated polybutadiene (HTPB) and hydroxy-carboxyl terminated polybutadiene (HCTPB), as well as their application to the encapsulation of capacitors.

EXPERIMENT

The epoxy resin used was E51, in which the epoxy value is 0.48-0.54. The molecular formulas of HTPB and HCTPB are $HO(CH_2-CH=CH-CH_2)_n OH$ and $HO(CH_2-CH=CH-CH_2)_n COOH$, respectively. Some of the properties of the blends which were used in our research work are listed in Table I.

Table I Some properties of HTPB and HCTPB

	M_n	-OH	-COOH	viscocity (40°C)	
HTPB	3850	0.5680		4.28	colourless thick liquid
HCTPB	2784	0.5176	0.4462		brownish thick liquid

The curing agent used was hardene-500.
The blending process was as follows: the liquid rubber was dissolved in the resins, then the curing agent was added with stirring, and was degassed by vacuum treatment. After that, the blends were cast into the sample mold followed by curing at the condition of 40°C/2hr + 60°C/2hr + 80°C/2hr + 120°C/2hr.
For the test of encapsulation of capacitors, the dipping method was used.
The impact strength of toughened epoxy resins was measured with an XJJ-500

impactor according to GBO43-79. The stretch strength was tested with a WJ-10A universal testing machine according to GB1040-79. The dynamic mechanical analysis was done on a DDV(II)-EA viscoelastometer.

The insulation performance of the encapsulated capacitors was measured by an LCR-4274A tester.

RESULTS AND DISCUSSION

HTPB and HCTPB are two kinds of telechelic polymers which are easily prepared. They are of great practical significance since they play an important role in toughening plastics.

Phase Separation

It can be seen from Fig. 1 and Fig. 2 that hydroxy-carboxyl rubber and hydroxyl-terminated polybutadiene have a remarkable toughening effect on the epoxy resin, which increases with the increase in rubber content. The necessary condition for obtaining optimum performance from the blends is the reasonable compatibility of the elastomer with the resin; i.e., dissolved rubber disperses uniformly in the resin and the phase separation can easily take place during solidification. The rubber particles, as stress concentrating agents, play a role in controlling cracking. The formation of two phases contributes to the improvement in toughness.

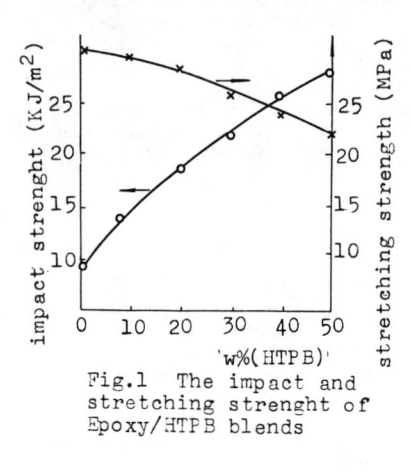

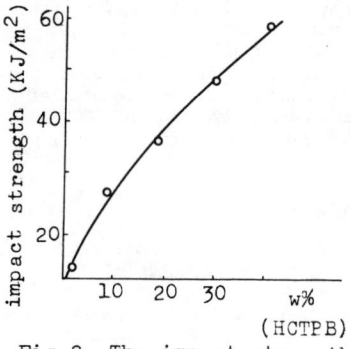

Fig.1 The impact and stretching strenght of Epoxy/HTPB blends

Fig.2 The impact strength of Epoxy/HCTPB blends

Fig. 3 is the SEM photo of HTPB blends and Fig. 4 is that of HCTPB.
It is clear that the dispersion of rubber particles is quite uniform and the phase domain is smaller in the HTPB blend, which shows that the phase separation process is quite efficient.

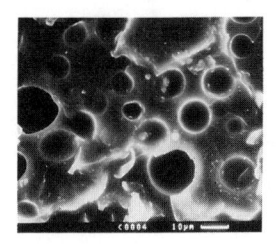

Fig. 3 The SEM photograph of epoxy/HTPB blends

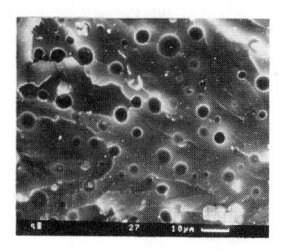

Fig. 4 The SEM photograph of epoxy/HCTPB blends

From a thermodynamic perspective, besides resin-rubber miscibility, the quality of phase separation also depends on the molecular weight of rubber. If the molecular weight is too high, the rubber dissolves in resins with difficulty, but if the molecular weight is too low, the phase separation cannot readily occur. It has been noted that when the molecular weight is about 3800, phase separation can occur and a high break strength can be obtained. The molecular weight of HTPB and HCTPB used in our research work is suitable.

Mechanical Transitions and Craze Resistance of Blending Systems

Fig. 5 indicates the curves of dynamic mechanical analysis for epoxy/HTPB blends. In Fig. 5 it can be seen that there is a transition peak at a low temperature of about -80°C, which is attributed to the main transition (glass transition) of HTPB, while the secondary transition (β transition) peak of epoxy resin appears near -40°C which occurs by some movement of short segments in the main chain. According to toughening theory, the lower the glass transition temperature of the rubber phase, the better the toughening effect. At the testing temperature and under impact loads applied at a high rate, only those blends containing particles with a sufficiently low T_g can achieve good toughness. If the T_g is higher and the rubber particles are in a glassy state, the relaxing deformation cannot take place; and mechanisms for limiting crack propagation, such as shear banding, are lost. Generally, when the T_g of rubber is about 60° lower than that of the test temperature, enough of a relaxing deformation can occur. The transition temperature of the blends is between -80°C and -40°C, giving the epoxy/HTPB excellent craze resistance.

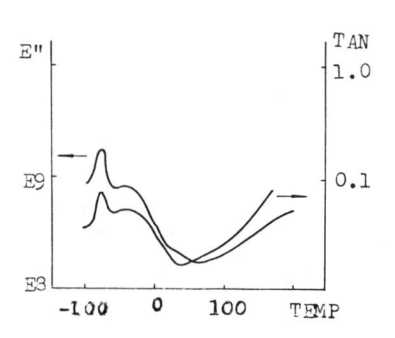

Fig. 5 DMA photograph for epoxy/HTPB

The Application to Encapsulation of Capacitors

The excellent craze resistance at low temperatures of epoxy resins modified by HTPB or HCTPB has been applied to the encapsulation of capacitors. One of the technical specifications for a ceramic capacitor encapsulated with epoxy resin is that no crazing occurs while it undergoes a -55°C to 80°C cyclic test 5 times. If the epoxy resin formulation does not contain the elastomer, it cannot pass the above test. Epoxy resins toughened with HTPB or HCTPB exhibit no crazing when tested as above, and they exhibit good electrical performance. Table II shows the experimental values of encapsulated capacitors.

Table II The electrical performances of encapsulated capacitors

Blends	Epoxy/HCTPB		Epoxy/HTPB	
Items	before cyclic test	after cyclic test	before cyclic test	after cyclic test
Endurance voltage KV	14	13.5	13	12.8
before wet heat test				
volume resistance.	1.6×10^{12}	1.3×10^{11}	10×10^{11}	9.3×10^{10}
dielectric loss	8.5×10^{-3}	8.9×10^{-3}	7.6×10^{-3}	8.3×10^{-3}
after wet heat test				
volume resistance.	2.8×10^{10}	2.1×10^{10}	6.6×10^{10}	5.3×10^{10}
dielectric loss	9.0×10^{-3}	9.3×10^{-3}	9.5×10^{-3}	9.7×10^{-3}

Notes:
1. tested sample: 3KV. 1000pf ceramic capacitor
2. technical specification
   endurance voltage: 7.5KV
   volume resistance:
       room temperature: 4×10^9
       after wet heat test: 2×10^9
   dielectric loss:
       room temperature: 3×10^{-3}
       after wet heat test: 6×10^{-3}

It can be seen from Table II that high-low temperature cycling has no great effect on the electrical properties of capacitors encapsulated with blends of epoxy/HTPB or epoxy/HCTPB. After wet heat testing, the decrease of volume resistance and increase of dielectric loss are also much smaller than required by the technical specification. These two telechelic polymers, as toughening agents, yield satisfactory results in the encapsulation of capacitors.

The Comparison of Toughening Effects Between HTPB and HCTPB

The molecular structural difference between HTPB and HCTPB is that the latter has a carboxyl group at one end. This carboxyl group can react with epoxy resin to form a block containing flexible segments. The flexible block enters the cured epoxy resin structure and makes the cross-linked epoxy have a good flexibility. There is no carboxyl group in the HTPB, and no chemical bond occurs between HTPB and epoxy. In this case, the toughening effect arises only from physical forces. The result is not as good as in HCTPB. Experiments showed that when epoxy/HCTPB=(100/20), the impact strength increases about 3.2 times in relation to the untoughened epoxy; but when HTPB/epoxy=(20/100), it only increases about 2 times (see Fig. 1 and 2). When the rubber content was increased, the impact strength of HCTPB blends increased remarkably. The toughening effect from the addition of HCTPB and HTPB also had an effect on the electric properties of the encapsulated capacitors (see Table II). The endurance voltage of epoxy/HCTPB is higher than that of epoxy/HTPB, offering remarkable evidence of the above conclusion.

CONCLUSION

HTPB and HCTPB toughen epoxy resins effectively. The impact strength increases with an increase in the content of telechelic polymer in blends.

Epoxy/HTPB and epoxy/HCTPB exhibit excellent craze resistance at low temperatures, due to a low T_g in the rubber phase. These blends have been applied to the encapsulation of capacitors and have solved the problem of crazing at -55°C.

HTPB-epoxy miscibility and HCTPB-epoxy miscibility are formidable, and the phase separation occurs gradually during solidification to form a two-phase structure. The existence of discrete rubber particles dispersed in the epoxy matrix gives the blend systems higher impact strength.

REFERENCES

(1) R.E. Cais, Macromolecules, 13, 415 (1980).
(2) E.H. Rowe, Mod. Plast., 49, 110 (Aug. 1970).
(3) Dexiu Huong and Zhongyuan Ren, J. Xian Jiao Univ., 3, 23 (1984).
(4) Zhongyuan Ren and Liying Qi, J. Thermosetting Resin, 2, 6 (1990).

IN-SITU PHASE SEPARATION OF AN AMINE-TERMINATED SILOXANE IN EPOXY MATRICES

D. F. Bergstrom, G. T. Burns, G. T. Decker, R. L. Durall, D. Fryrear, G. A. Gornowicz, Dow Corning Corporation, Midland, MI 48686-0994
M. Tokunoh, N. Odagiri, Composite Materials Research Laboratories, Toray Industries, Inc. Iyogun, Ehime 791-31 Japan

ABSTRACT

We have developed an amino-functional silicone resin to toughen epoxies which, when prereacted with the epoxy function in resins, undergoes in-situ phase separation during final epoxy curing. SEM analyses of the morphology of fracture surfaces of MY720-DDS, Epon 828-DDS and other epoxy matrices modified with the silicone resin showed rough surfaces with the formation of very uniform <10 μm spheres. Silicon and sulfur elemental distribution mapping showed silicon rich spheres embedded in an epoxy matrix. We report cavitation, particle debonding and pull-out, and an increases in fracture surface area as possible modes of toughening. Silicone modified materials give improvements in slow strain rate G_{1c} fracture toughness measurements of 250-400%, similar to carboxy terminated polybutadiene-acrylonitrile copolymer (CTBN) modifiers, but with a much smaller flexural modulus loss. The T_g of the modified epoxy matrices are maintained, moisture resistance is improved, and flammability is reduced.

INTRODUCTION

Cured, highly crosslinked epoxy resins provide superior adhesion and strength in composites and laminates but continue to show poor damage tolerance due to brittle fracture. Two methods of epoxy toughening are optimization of the crosslink density [1] and incorporation of second phases [2] to form damage-tolerant matrices. Unfortunately, these methods of toughening often sacrifice matrix modulus and Tg. Discrete second phases are typically added by physical blending or chemically grafting an incompatible polymer into the uncured epoxy matrix. Examples of blended nonreactive thermoplastic modifiers are nonreactive polyether sulfones and polyether imides [3]. Blending in nonreactive thermoplastic modifiers can effectively toughen epoxies, however, chemically grafting an incompatible polymer to an epoxy matrix polymer usually results in stronger adhesion of the modifier particles to the epoxy matrix and generally leads to fewer processing problems [4]. Grafted reactive modifiers which undergo in-situ phase separation during epoxy curing are liquid polybutadiene-acrylonitrile copolymers [4], liquid linear siloxane oligomers [5], and reactive amino-functional polyether sulfones and polyether ketones [6]. However, with grafted systems problems with loss of flexural modulus and T_g persist.

Siloxanes, known for superior thermal and oxidative stability and hydrophobicity, have been used for epoxy modification [5]. These have included amine-terminated polydimethylsiloxanes modified with methyltrifluoropropyl- or diphenyl- siloxane units. The problems of loss of modulus and T_g have persisted with siloxanes containing predominantly the soft, flexible dimethylsilyl group. We have developed a higher modulus, amino-functional phenyl silicone resin for epoxy modification which is compatible with uncured epoxy resins but phase separates from the epoxy matrix during curing [7]. We have toughened MY 720 cured with diaminodiphenyl sulfone (DDS), Epon 828-DDS, and other epoxy matrices with the new amino-functional silicone. We have found substantially less loss of matrix flexural modulus upon modification with the new siloxane resin than with CTBN. The T_g of the epoxy was maintained and moisture resistance was improved. Addition of the siloxane phase reduces the heat release rate of the matrix. Plasma oxidation of polished, unfractured surfaces indicates good adhesion of the siloxane modifier particles to the matrix. However, fractured surfaces show particle cavitation, debonding and pull-out as possible fracture energy absorbing processes.

EXPERIMENTAL

The amine-terminated siloxane resin for epoxy modification is a phenylated, amine functional silicone resin prepared by standard equilibration techniques from phenyl, diphenyl and aminopropyl siloxane precursors. The molecular weight of the resin product can be varied by changing processing conditions with typical values being M_n= 1300 and M_w= 2200. Amine equivalent weight is typically found to be 1170 g/eq.

To modify the epoxies with the silicone resin, the epoxy resin and the amino-functional siloxane resin and toluene sufficient for dissolution were mixed, prereacted, devolatilized and cooled before crosslinkers were added. The modified epoxy/crosslinker mixture was then heated and poured into preheated molds and cured. A more detailed explanation of sample preparation is contained elsewhere [7].

Fracture toughness was defined by the measurement of G_{Ic}, the critical strain energy release rate [7]. G_{Ic} values are from the compliance calibration technique using a double torsion test geometry [8].

Flexural properties of samples were tested according to the ASTM Standard D790-90 three point loading system. A flexural modulus was calculated from the slope of the load displacement curve at a fixed strain of 0.5 mm.

Micrographs were taken of the epoxy fracture surfaces using a JOEL T300 Scanning Electron Microscope at 15 KeV accelerating voltage. Samples were mapped for silicon using a Cameca MBX Scanning Electron Microprobe at 15 KeV accelerating voltage.

To examine the morphology of unfractured surfaces, unfractured samples were first polished and then carefully plasma oxidized in a Branson/IPC Plasma Chamber at 100 watts for 10 minutes at 0.3 torr until the spheres just appeared but no etching into particle/matrix interface was done.

Flammability was done using a cone calorimeter.

RESULTS

The silicone oligomer is prereacted by heating with the epoxy resin, prior to curing, forming a homogeneous silicone-epoxy grafted copolymer. As the epoxy is then crosslinked, the copolymer becomes incompatible with the epoxy network and phase separation occurs. Scanning electron micrographs (SEM) of fracture surfaces of modified MY 720-DDS and Epon 828-DDS epoxy samples show the presence of small uniform <10 µm spheres as a second phase (Figure 1). Unmodified samples are essentially featureless. Electron microprobe results showed the spheres to be rich in Si and the continuous phase to be richer in sulfur [7]. Thus the spheres are a silicon-rich epoxy phase and are surrounded by an epoxy matrix. Cavitation, particle debonding and pull-out are seen on fracture surfaces. These are possible fracture energy absorbing events. To verify that debonding results from fracturing and not from different thermal coefficients of expansion during curing, unfractured sample surfaces were polished and then plasma oxidized just until the particles appeared. The resulting scanning electron micrograph (Figure 2) shows the spheres with good adhesion to the epoxy matrix after processing but prior to fracturing, indicating that

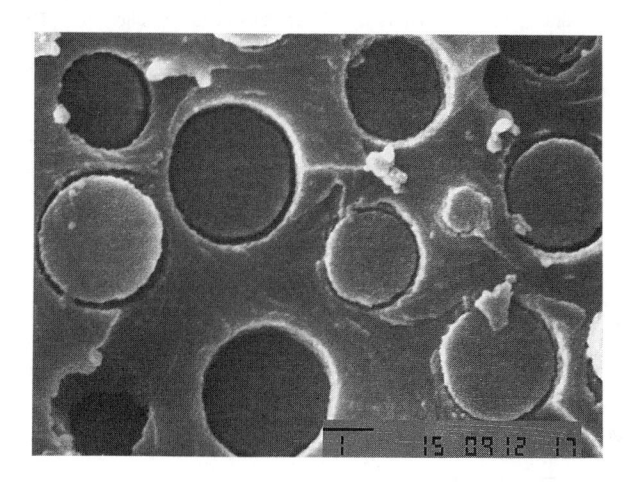

Figure 1. Fracture Surface Epon 828-DDS, 20% Siloxane, 10,000x (length of bar is 1 μm)

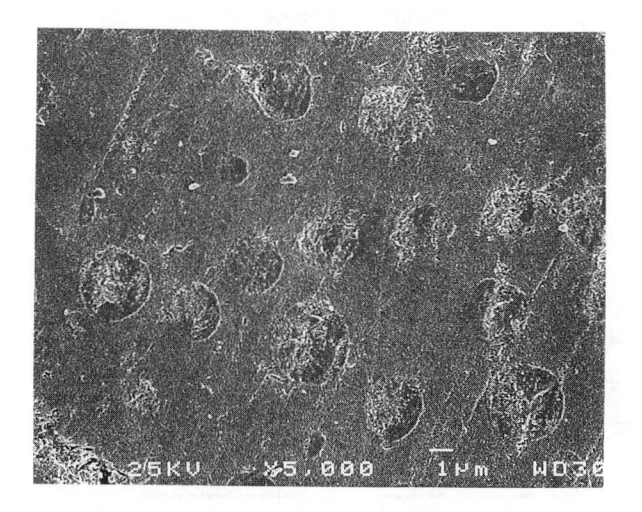

Figure 2. Plasma Etched Surface of Epon 828-DDS, 20% Siloxane, 5000x (length of bar is 1 μm)

the debonding is indeed occurring during fracturing.

We established an increase in fracture toughness with a maintenance of flexural properties. Figure 3 shows a plot of G_{Ic} versus weight percent additive for samples of MY720-DDS containing various amounts of amino-functional silicone. The silicone increased the fracture toughness of MY 720-DDS by 2.5 fold at 20% loading. At 10 % by weight modifier, the amino-functional siloxane and CTBN gave comparable toughening in MY 720-DDS. However the CTBN caused a much more substantial loss of modulus (Figure 4).

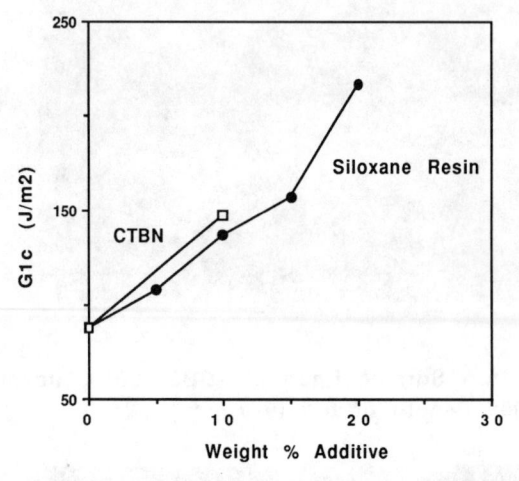

Figure 3. Fracture Toughness of Modified MY 720-DDS

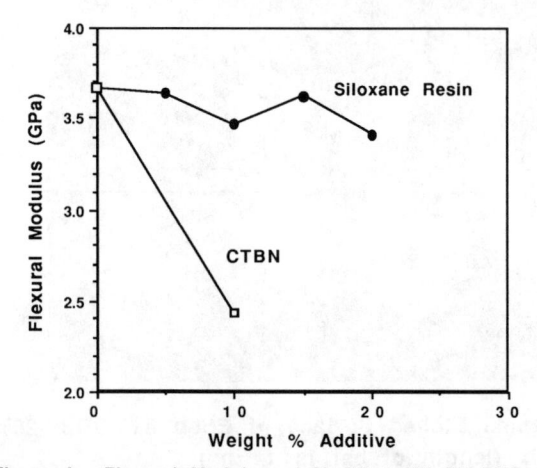

Figure 4. Flexural Modulus of Modified MY 720-DDS

The silicone resin also toughens Epon 828 cured with DDS. We have noted a superior 4.25 fold increase in G_{Ic} for Epon 828-DDS with 20% silicone resin. The better increase in toughness in the Epon 828 sample may be a result of smaller, more evenly distributed particles which were found in the separated phase of the modified Epon 828-DDS sample.

The improvements in G_{Ic} were achieved without significant losses in T_g. The T_g of a sample of MY 720-DDS before and after modification with the siloxane resin were found to be 235 °C and 237 °C respectively. Similarly, The T_g of Epon 828-DDS before and after modification were found to be 201 °C and 202 °C respectively [7].

Cone calorimetry data (Figure 5) shows improved heat release rates of Epon 828-DDS epoxy modified with only 20 % by weight of the siloxane at heat fluxes of both 30 and 60 kilowatts/m^2. There is however, a slightly reduced time to ignition with siloxane modification. A comparison of the heat released at 30 kilowatts/m^2 between epoxy samples modified with two different lots of the siloxane resin showed reasonable agreement.

The amino-terminated siloxane resin is one of a variety of resin compositions investigated for toughening MY 720-DDS and Epon 828-DDS. The composition and molecular weight of our matrix modifier resins are easily varied with our preparative procedure. The effects of variations in the molecular weight of the siloxane resin on its performance are currently under investigation.

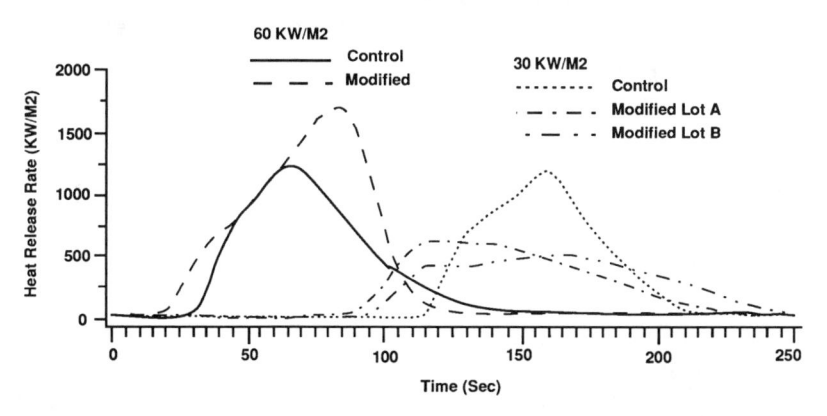

Figure 5. Cone Calorimetry Data for Epon 828-DDS with 20 weight % siloxane. Lower heat release at both 30 and 60 KW/m^2 fluxes.

CONCLUSIONS

Our reactive amino-functional silicone resin can effectively toughen MY 720-DDS and Epon 828-DDS epoxies by 250-425% at very little cost to the matrix modulus and T_g in comparison to CTBN modifier. Interesting in-situ phase separation occurs during epoxy curing to form uniform <10 μm spheres of a silicon-rich phase embedded in an epoxy matrix. The spheres appear to be well bonded to the matrix in unfractured epoxy surfaces but undergo cavitation or debonding and particle pull-out during fracturing. These energy absorbing processes and the subsequent formation of larger surface areas on the fracture surface are likely sources of the increases in fracture toughness. We have observed improvements in heat release rates for the modified epoxies as well.

REFERENCES

1. G. Levita, in *Rubber-Modified Thermoset Resins: Advances in Chemistry 208*, edited by C. K. Riew and J. K. Gillham (American Chemical Society, Washington, DC, 1984) pp. 93-118.
2. W. D. Bascom, D. L. Hunston, in *Rubber-Modified Thermoset Resins: Advances in Chemistry 208*, edited by C. K. Riew and J. K. Gillham (American Chemical Society, Washington, DC, 1984) pp. 135-172.
3. (a) C. B. Bucknall, I. K. Patridge, Polymer, 24, 639 (1983); (b) J. Daimont, R. J. Moulton, Science of Advanced Materials and Process Engineering Series 29, 422 (1984).
4. (a) E. H. Rowe, A. R. Siebert, R. S. Drake, Mod. Plast., 47, 110 (1970); (b) J. M. Sultan, F. McGarry, Polym. Eng. Sci., 13, 29 (1973).
5. J. S. Riffle, I. Yilgor, A. K. Banthia, C. Tran, G. L. Wilkes, J. E. McGrath, J. Am. Chem. Soc. Symp. Ser., 221, 21 (1982).
6. J. A. Cecere, J. L. Hedrick, J. E. McGrath, Science of Advanced Materials and Process Engineering Series, 31, 580 (1986).
7. D. F. Bergstrom, G. T. Burns, G. T. Decker, R. L. Durall, D. Fryrear, G. A. Gornowicz, M. Tokunoh, N. Odagiri, Science of Advanced Materials and Process Engineering Series, 37, 278 (1992).
8. S. M. Lee, J. Mat. Sci. Lett., 511 (1982).

HIGH TEMPERATURE POLYMER NANOFOAMS

J. HEDRICK, J. LABADIE, T. RUSSELL, V. WAKHARKAR and D. HOFER
IBM Almaden Research Center, 650 Harry Road, San Jose, California 95120-6099

ABSTRACT

A means of generating high temperature polymer foams which leads to pore sizes in the nanometer regime has been developed. Foams were prepared by casting block copolymers comprised of a thermally stable block and a thermally labile material, such that the morphology provides a matrix of the thermally stable material with the thermally labile material as the dispersed phase. Upon a thermal treatment, the thermally unstable block undergoes thermolysis leaving pores where the size and shape of the pores are dictated by the initial copolymer morphology. Nanopore foam formation is shown for triblock copolymers comprised of a poly(phenylquinoxaline) matrix with poly(propylene oxide) as the thermally labile block. Upon decomposition of this block, a 10-20% reduction in density was observed, consistent with the initial PO composition, and the resulting PPQ foam showed a dielectric constant of ~ 2.4, substantially lower than PPQ (2.8). Small angle X-ray scattering and transmission electron microscopy show pore sizes of approximately 100Å.

INTRODUCTION

Polymeric materials have become increasingly important as interlayer dielectrics, passivation layers and structural resins. These insulators must withstand severe thermal, chemical and mechanical stresses associated with microelectronics fabrication, and, in such situations, polymeric materials have many shortcomings when compared to the inorganic alternatives such as alumina or silicon oxides. However, polymeric insulators have several key attractive features including low dielectric constant and ease of processing. Since the velocity of pulse propagation is inversely proportional to the square root of the dielectric constant of the medium, reductions in the dielectric constant of the insulating material, translate directly into improvements in machine cycle time [1]. Furthermore, the minimum distance between lines is dictated by cross-talk or the noise that results from induced current in conductors adjacent to active signal lines. This cross-talk is directly dependent upon the dielectric constant of the insulating material [1]. A reduction in the dielectric constant of the insulator allows the signal lines to be closer, further improving machine cycle time.

Although polyimides meet most of the material requirements for microelectronics applications, it would be desirable to improve the electrical properties of these materials by reducing the dielectric constant and water absorption. The most common approach to modify the dielectric properties of polyimides has been through incorporation of perfluoroalkyl groups. Examples include the incorporation of hexafluoroisopropylidine linkages (Hoechst Sixef) [2], main chain perfluoroalkylene groups [3], and pendant trifluoromethyl groups [4,5]. These approaches afford materials with dielectric constants below 3.0 (as low as 2.6 for Hoechst Sixef) and low water absorption ($\simeq 0.5\%$). Unfortunately, modifying the polymer by introducing flexible perfluoroalkyl groups in the backbone produces a material with diminished final properties. The addition of pendant trifluoromethyl groups appears to have a less negative affect on the final properties, but is limited by the synthetic methodology required to incorporate enough trifluoromethyl groups to have an affect on the dielectric constant.

An alternative approach to reducing the dielectric constant substantially while maintaining the desired thermal and mechanical properties of the aromatic polyimide is to generate a polyimide foam. The reduction in the dielectric constant is simply achieved by replacing the polymer, having a dielectric constant ~ 3.2, with air which has a dielectric constant of 1. However, the criteria for the size of the pores or voids is stringent. It is mandatory that the size of the voids be much smaller than the film thickness and/or the microelectronic features for the gain in the dielectric constant to be realized. Secondly, it is necessary that the pores be closed cell, i.e., the connectivity between the pores be minimal. A substantially open celled structure would not necessarily hamper the dielectric performance of the film. However, in an actual application the rate at which solvent from subsequent layers or other impurities would be incorporated within the film would be high. Finally, the size of the pores, the extent to which the pores are interconnected and the volume fraction of the pores can alter the mechanical properties and structural stability of the foamed material. If any of these is too large, then the foam can undergo a collapse which essentially regenerates the original material and, hence, no improvement in the dielectric constant. Many of the high temperature polymer foams reported to date fail in one of these categories or the matrix polymer does not have the thermal stability required for application in microelectronics and, hence, have not found any practical use.

Herein, the synthesis and properties of a poly(phenylquinoxaline), PPQ, foamed polymer are reported. As will be shown, the foam is prepared by the synthesis of a PPQ copolymer with a coblock comprised of a polymer which undergoes a thermal decomposition to volatile products. The use of a block copolymer generates a two phase morphology with the thermally unstable polymer comprising the minor phase in a matrix of the PPQ. Upon degradation of this component, voids are left in a matrix of the PPQ with a size commensurate with that of the precursor copolymer. By adjusting the molecular weights of the two components, the volume fraction and size of the resultant voids can be varied. For the initial experiments reported in this article, triblock copolymers of PPQ with poly(propylene oxide), PO, were used where the latter undergoes thermolysis to volatile products and, hence, are easily removed.

EXPERIMENTAL

Materials

Bis(phenylglyoxalyl)benzene (BPGB), 3,3'-diaminobenzidine (DAB), and 4-hydroxybenzil were obtained and purified as described previously [6]. Monofunctional poly(propylene oxide) (PO) oligomers with Mn of 5,000 and 10,000 were kindly provided by Dow Chemical and were kept *in vacuo* prior to use. The 1.9 M phosgene/toluene solution was obtained from Fluka and was used as received.

Benzil End-Capped PO Oligomers

The general method for functionalization of PO oligomers with a benzil end group is given for the 5K oligomer. A 100 mL three-neck round bottom flask fitted with a dry ice condenser and a nitrogen inlet was charged with 10.0 g (2.00 mmole) of PO followed by 25 mL (50 mmole) of phosgene/toluene solution. The stirred mixture was heated to 60 °C for 2 h in a water bath. The phosgene and toluene were removed with a strong stream of nitrogen, followed by evacuation for several hours. The resulting viscous oil was diluted with 50 mL of methylene chloride and 430 mg (1.90 mmole) of 4-hydroxybenzil was added. The mixture was cooled to 0 °C and 5 mL of pyridine was added dropwise. The reaction mixture was allowed to warm to room temperature overnight and then partitioned between methylene chloride and 5% aqueous HCl. The organic layer was separated and washed twice with 5% aqueous HCl, water and dried with magnesium sulfate. Concentration of the organic extract on a rotary evaporator, followed by pumping 18 h at 0.1 mm Hg afforded 8.2 g of the desired oligomer as a clear oil. Comparison of the benzil and PO backbone spectral resonances in the H NMR showed the oligomer to have an Mn = 6,000.

The 10K PO oligomer was prepared by the general procedure using 10 g (1 mmole) of monofunctional hydroxy end-capped PO, 25 mL of phosgene in toluene, 215 mg (0.95 mmole) of hydroxybenzil, 2.5 mL pyridine, and 50 mL of chloroform. Analysis by H NMR showed the oligomer to have a molecular weight of 9400.

Poly(phenylquinoxaline)-PO Block Copolymers

Block copolymers were prepared under conditions analogous to conventional PPQ synthesis [7-9]. A copolymer with a 20% theoretical composition of the 6K PO was synthesized by slurrying 7.1517 g (33.377 mmole) of DAB in 40 mL of m-cresol/xylene (1:1). A solution of 11.313 g (33.050 mmole) of BPGB and 4.00 g (0.66 mmole) of the benzil end-capped oligomer in 100 mL of m-cresol/xylene were added to the slurry dropwise with stirring. When the addition was complete, the addition funnel was rinsed with 30 mL of m-cresol/xylene and the polymerization mixture was stirred for 20 h. The polymer was isolated by precipitation in methanol followed by three methanol washes. The copolymer was isolated in 88% yield. Analysis of the copolymer by H NMR showed 13.4 wt% PO composition by integration of aromatic PPQ resonances vs. the PO resonances. The analogous copolymer based on the 9.4K oligomer was isolated in 90% yield and had a PO composition of 16.5 wt%.

Foam Formation

The copolymers were dissolved in tetrachloroethane (TCE) at a concentration of 9% solids. Coatings of 10 μm thickness were obtained by spin coating at 2000 rpm on 2.54 cm diameter low resistivity Si wafers. The removal of the solvent TCE was accomplished by heating the polymer films to 150 °C at 5 °C/min and maintaining them at 150 °C for 2 hours in a nitrogen atmosphere. The films were then cooled to room temperature and reheated to 275 °C at a rate of 5 °C/min and maintained at 275 °C for 9 hours in air.

Measurements

The dielectric measurements on the polymer films were carried out using a HP multifrequency analyzer. Medium frequency (10 kHz to 4 MHz) capacitance and loss tangents were measured in the temperature range of 25-100 °C. The specimens for the dielectric measurements were prepared by vacuum deposition of 3 mm diameter gold electrodes on the spin coated polymer films with the back surface of the low resistivity Si wafer serving as the other electrode. Glass transition temperatures, taken as the midpoint of the change in slope of the baseline, were measured on a DuPont DSC 1090 instrument with a heating rate of 10 °C/min. The dynamic mechanical measurements were performed on a Polymer Laboratories Dynamic Mechanical Thermal Analyzer (DMTA) at 10 Hz and a heating rate of 10 °C/min in the tension mode. Isothermal and variable temperature (5 °C heating rate) thermal travimetric analysis (TGA) measurements were performed on a Perkin-Elmer model TGA-7 in a nitrogen atmosphere. Density measurements were obtained with a density gradient column made from carbon tetrachloride and xylene. The column was calibrated against a set of beads of known densities at 25 °C. At least three specimens were used for each density measurement.

Small angle x-ray scattering (SAXS) measurements were performed on a Kratky camera. Approximately 0.5 meter from the specimen was mounted a TEC 210 position sensitive proportional counter which served to collect the entire scattering profile simultaneously. The scattering profiles were accumulated on a multichannel analyzer after pulse height discrimination. Ni filtered Cu radiation from a Rigaku 18 kW rotating anode generator operated at 40 Kv and 100 mA was used for these studies.

RESULTS AND DISCUSSION

PPQ was chosen as the matrix polymer due to its high thermal stability and glass transition temperature, Tg, and its good processability. PO was used as the thermally decomposable block due to its immiscibility with PPQ, thermal lability at 250 °C, and its availability in the form of hydroxyl functionalized oligomers. To maximize the size and purity of the PO domains, PO-PPQ-PO triblock copolymers with relatively high molecular weight PO blocks were prepared. In addition, for a given composition, using a higher molecular weight PO block will lead to a copolymer comprised of a higher molecular weight PPQ block as well, benefitting mechanical properties of the foam.

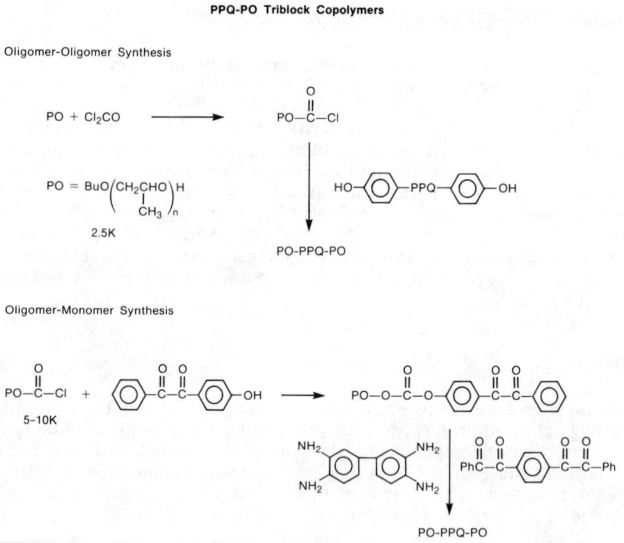

PPQ-PO Triblock Copolymers

Scheme I

The synthesis of the triblock copolymers was carried out by a monomers-oligomer approach (Scheme 1). The triblock copolymers were prepared by addition of a mixture of BPGB and a benzil end-capped PO oligomer to DAB. The benzil end-capped PO oligomers were prepared by converting monofunctionally terminated hydroxy PO to the chloroformate with phosgene, followed by condensation with hydroxybenzil. The procedure was carried out with 5K and 10K PO oligomers to give benzil end-capped PO with Mn's of 6,000 and 9,400, respectively. The benzil end-capped PO oligomer were subsequently used in the synthesis of PPQ-PO triblock copolymers with a target PO composition of 20 wt% by adjusting the stoichiometry of the PO, BPGB and DAB according to the Carothers' equation (Table I).

The compositional and thermal characteristics of the phenylquinoxaline-propylene oxide copolymers are shown in Table I. For the initial experiments, the PO composition was maintained below 20 wt% with the objective of having discrete and noncontinuous domains of PO in a PPQ matrix to facilitate the formation of a stable foam upon PO decomposition. The copolymer composition was determined by H NMR by comparison of the integration of the PPQ and PO resonances. The PO compositions for triblock copolymers derived from the 6K (copolymer 1) and 9.4K (copolymer 2) oligomers were 13.4 and 16.5 wt%, respectively (Table I). The volume fraction PO in copolymers 1 and 2 was 16.6 and 20.2%, based on specific gravities of 1.028 for PO and 1.32 for PPQ (Table I). The theoretical Mn of the PPQ

Table I

Characteristics of Phenylquinoxaline-Propylene Oxide Copolymers

Specimen	$(M_n)_{PO}$ (g/mol)	Weight % PO Theoretical	Weight % PO Experimental	Volume Fraction %	Tg (°C)
PPQ	—	—	—	—	—
Copolymer 1	6,000	20	13.4	16.6	−45,360
Copolymer 2	9,500	20	16.5	20.2	−42,361

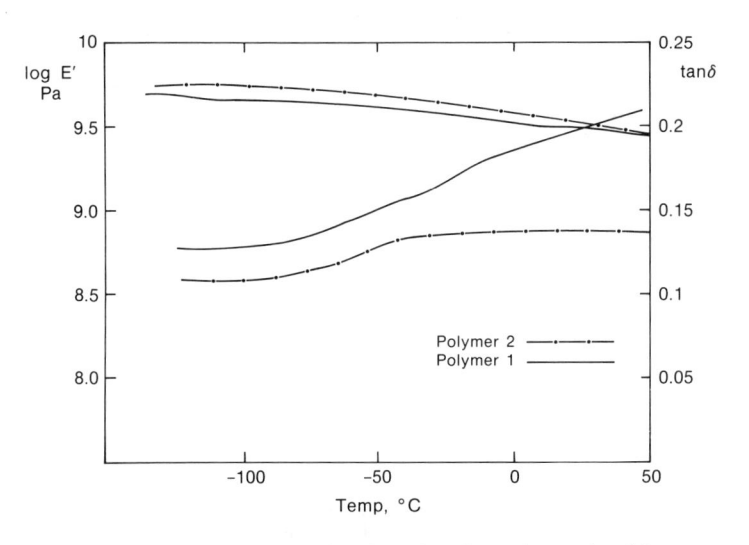

Figure 1. Dynamic mechanical behavior of Copolymers 1 and 2.

blocks, as determined by the stoichiometric offset used, was 48,000 and 75,000 for the copolymers derived from the 6K and 9.4K PO, respectively. Due to the high PPQ block molecular weight required to give a triblock copolymer, it is likely that some PPQ-PO diblock copolymers are present. For each of the copolymers, two Tg's were observed, ~ −40 and ~360 °C (Table I), indicative of a microphase separated morphology. The Tg of the PO in the copolymer was somewhat higher than the initial oligomer suggesting at least a partial phase mixing in this domain. Conversely, the PPQ transition in the copolymer was identical to that of the homopolymer, consistent with the high molecular weight of this block. The dynamic mechanical measurements, consistent with calorimetric results, show two transitions, and the low temperature regime for the two copolymers are shown in Figure 1.

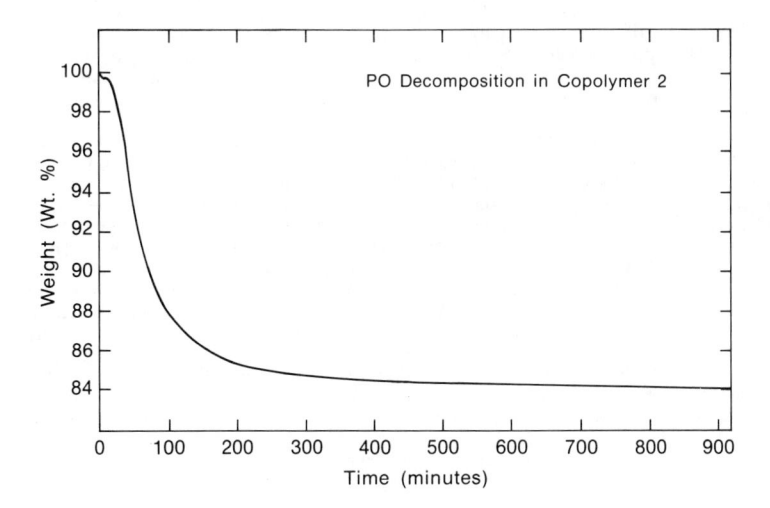

Figure 2. Thermogravimetric analysis of Copolymer 2.

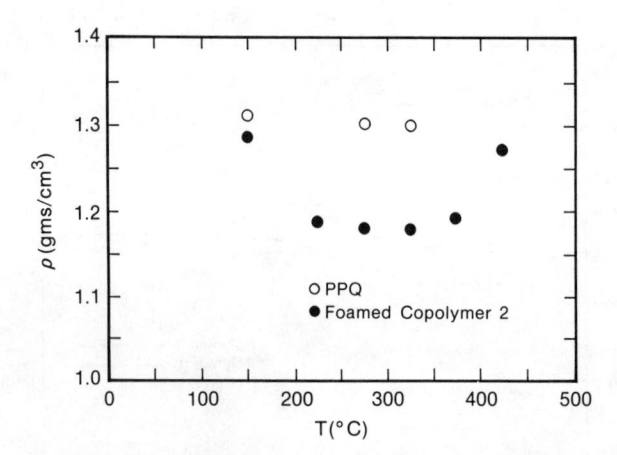

Figure 3. Density measurements as a function of temperature for PPQ and Copolymer 2.

A key component of generating a nanofoam is the ability to process the block copolymer foam precursor by conventional methods. The PO decomposition is high enough to allow removal of common solvents prior to PO degradation, and after solvent removal the PO can be decomposed with a subsequent heat treatment. The copolymers were dissolved in TCE and spun into thin films (~10 μm). The films were then heated to 150 °C for 2 h to remove the solvent. The films were then heated to 275 °C (9 h) in air to thermally decompose the PO. Isothermal gravimetric analysis indicated that the decomposition of the PO in the copolymer was complete under these conditions (Figure 2).

The density of the polymer clearly shows the formation of a foamed polymer. Shown in Figure 3 are the densities of a PPQ homopolymer and of a PO-PPQ-PO triblock copolymer as a function of process temperature. For the pure PPQ the density measured at room temperature was essentially independent of process temperature at 1.32 g/cm^3. The PO-PPQ-PO triblock copolymer, on the other hand, underwent substantial changes. After drying at 150 °C the density was 1.29 g/cm^3, consistent with volume additivity of the PO and PPQ blocks. After heating to 200 °C the density dropped precipitously to 1.17 g/cm^3, which was 88% of the density of PPQ. Without further analysis, these data show unequivocally that 12% of the film is occupied by voids. The density remained at this low value until ~350 °C whereupon the density began to increase. At 400 °C, the density was quite close to that of the PPQ homopolymer indicating that the foam structure had collapsed. It is remarkable, however, that over a large range in temperature, the foam structure was maintained.

While the density showed that the generation of a foam had been successfully achieved, no information was provided on the size scale of the pores. Shown in Figure 4 is a TEM of a typical microtomed section of a PO-PPQ-PO after decomposition of the PO block. The dark regions in the micrograph correspond to the PPQ matrix. What is seen is that a porous structure is obtained with pores having an average cross section from 8–10 nm. From the cross section shown, it appears as if there is not a substantial amount of interconnection between the pores and the desired morphology has been achieved.

SAXS offers an alternate means of addressing this issue. Shown in Figure 5 are a series of SAXS profiles for the PPQ homopolymer, a PO-PPQ-PO copolymer and the copolymer after decomposition of the PO block. The data are plotted as the logarithm of the scattering as a function of the scattering vector, $Q = (4\pi/\lambda)\sin\theta$ where λ is the wavelength (1.542Å) and 2θ is the scattering angle. The scattering arising from the PPQ and the PO-PPQ-PO copolymer is weak. Any angular dependence to the scattering at high Q can be attributed to the weak nature of the scattering and the difficulty associated with properly subtracting background. The copolymer shows slight increase in the scattering at low Q which can be attributed to the two phase nature of the copolymer.

Figure 4. TEM micrograph of Copolymer 2.

Upon decomposition of the PO block nearly an order of magnitude increase in the scattering occurs. This is consistent with the decomposition of the PO and removal of the monomer to form a void. This corresponds to a dramatic enhancement in the electron density difference between the pores and the matrix and, hence, the large increase in scattering.

From the angular dependence of the scattering a measure of the size of the pores can be obtained. The type of morphology being dealt with is the classic structure studied by Debye et al. [10,11] where one has a dense two phase system with sharp boundaries where the matrix is a glass and the second phase is void. Here, a correlation length, a, can be defined as the average size scale of heterogeneities and the observed scattering is given by

$$I(Q) = I_e(Q)V\overline{(\Delta\rho)^2} \int_0^\infty \gamma(r)4\pi r^2 \sin Qr \, dr \qquad (1)$$

where $\gamma(r)$ is the correlation function at a distance r, $I_e(Q)$ is the Thomson scattering factor, $\overline{(\Delta\rho)^2}$ is the mean square electron density difference and V is the scattering volume. As shown by Debye et al. [10,11] $\gamma(r)$ for a random two phase system can be well approximated by $e^{-r/a}$. The correlation length, as shown by Kratky [12], is related to the average chord length or dimension of the phases by

$$\ell_1 = \frac{a}{\phi_2} \quad \text{and} \quad \ell_2 = \frac{a}{\phi_1} \qquad (2)$$

where ℓ_i is the chord length of phase i with volume fraction ϕ_i.

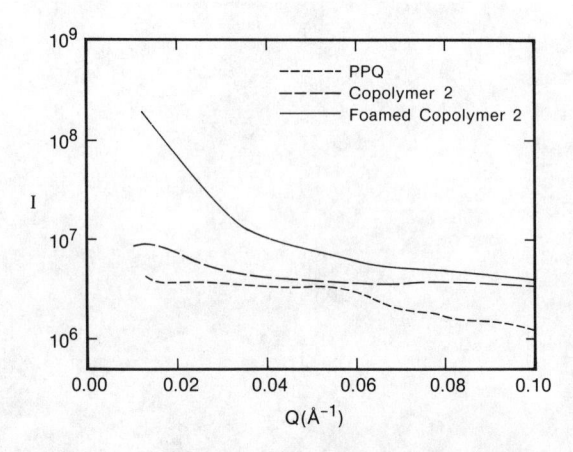

Figure 5. SAXS profiles for PPQ, Copolymer 2 and foam.

Performing this analysis on the PO-PPQ-PO triblock copolymer, after decomposition of the PO block, yielded the data shown in Table II. Assuming that the volume fraction of voids is given simply by the ratio of the density of the foamed copolymer divided by that of the PPQ, then correlation lengths of 8-9 nm are obtained which translates into chord lengths or average pore sizes of 8-8.9 nm. This is consistent with the TEM results shown previously and demonstrates the formation of pores on the tens of nanometer size scale.

The results obtained from the dielectric measurements are shown in Table II. The dielectric constant for the foamed copolymer 2 was 2.31 while that for PPQ homopolymer under similar conditions was 2.70. As compared to the PPQ homopolymer, this represents a reduction in the dielectric constant by almost 15% for the foamed copolymer.

SUMMARY

A new approach to low dielectric high temperature polymers has been demonstrated based on the generation of a foam of the desired high temperature polymer. These foams were prepared from block copolymers derived from a thermally stable block comprising the matrix and a thermally labile material. Upon a thermal treatment, the thermally unstable block decomposed back to monomer leaving pores of size and shape dictated by the copolymer morphology. Foam formation was demonstrated from triblock polymers of poly(phenylquinoxaline) and poly(propylene oxide) where the poly(propylene oxide) phase was decomposed as 275 °C leaving pores ~ 10 nm in size. The resulting foams showed the appropriate reduction in density and dielectric constant. This represents one of the first examples of high temperature polymer foams having pore sizes in the nanometer regime.

Table II

Characteristics of Poly(phenylquinoxiline) Nanofoams

Sample	$\phi_v^\ddagger$	Density, (g/cm^3)	a^+ (nm)	ℓ_p^* (nm)	Dielectric Constant, ε 45 °C, 2MHz
PPQ	—	1.32	—	—	2.70
Polymer 1	12.0	1.18	7.0	8.0	—
Polymer 2	9.8	1.16	8.1	8.8	2.31

REFERENCES

1. R.R. Tummala and E.J. Rymaszewski, Microelectronics Packaging Handbook (Van Nostrand Reinhold, New York, 1989a), Chapter 1.
2. Sixef(TM) is a commercial polyimide marketed by Hoechst-Celanese.
3. J.S. Critchlen, P.A. Gratan, M.A. White, and J.S. Pippett, J. Polym. Sci.: Part A-1 10, 1789 (1972).
4. F. W. Harris, S.L.C. Hsu, C.J. Lee, B.S. Lee, F. Arnold, and S.Z.D. Cheng, Mats. Res. Soc. Symp. Proc. 227, 3 (1991).
5. S. Sasaki, T. Matsuora, S. Nishi, and S. Ando, Mats. Res. Soc. Symp. Proc. 227, 49 (1991).
6. J. Labadie and J. Hedrick, J. Polym. Sci., Polym. Chem. Ed. 000 (1992).
7. P.M. Hergenrother and H.H. Levine, J. Polym. Sci., Polym. Chem. Ed. 5, 1453 (1967).
8. P.M. Hergenrother, J. Macromol. Sci., Rev. Macromol. Chem. C6, 1 (1971).
9. P.M. Hergenrother, J. Appl. Polym. Sci. 18, 1779 (1974).
10. P. Debye and A.M. Beuche, J. Appl. Phys. 20, 518 (1940).
11. P. Debye, H.R. Anderson, and H. Brumberger, J. Appl. Phys. 28, 679 (1957).
12. O. Kratky, J. Pure and Appl. Chem. 12, 483 (1966).

LASER OR FLOOD EXPOSURE GENERATED ELECTRICALLY CONDUCTING PATTERNS IN POLYMERS

JOACHIM BARGON AND REINHARD BAUMANN*
University of Bonn, Institute of Physical Chemistry, Wegelerstrasse 12,
D-W-5300 BONN-1, Germany, and
*Institute of Technology Leipzig, Fachbereich Naturwissenschaften, P.O. Box 66,
D-O-7030 Leipzig, Germany

ABSTRACT

Electrically conducting patterns can be generated in insulating polymers or composites either via UV-flood exposure through a mask or via laser irradiation. Various lithographic concepts starting either from conventional or custom tailored polymers or from special composites have been developed and tested. Thereby electrically conducting polymers are photogenerated either directly from a self-developing photosensitive precursor or via a two-component redox approach using one of the components as a vapor in an otherwise dry process. The electrically conducting patterns so obtained may be reinforced by plating them with metals electrogalvanically. These processes may also be combined with laser induced ablation, whereby the intensity of the laser beam may be gated to either induce electrical conductivity of the substrate or to ablate it without rendering it conductive. Analogously, thin films of electrically conducting polymers on top of insulating polymer layers can be patterned directly using excimer laser ablation.

INTRODUCTION

Polymers are typically known to be excellent electrical insulators, some of which, in particular the polyimides, have even found application on the chip level of microelectric circuitry. Others serve as indispensable photo-, e-beam or x-ray resists providing polymer patterns with a resolution in the micron or submicron range which serve to copy circuitry layouts into underlying silicon material.

Electrically conducting polymers have been known to exists for decades , but not until the discovery of electrically conducting heterocycles (i.e. derivatives of both pyrrole and thiophene) which excel due to their higher stability with respect to ambient atmosphere have these new materials found any significant industrial application. Recently from polymeric conductors, attractive devices were made, especially polymeric thin film field effect transistors [1,2,3], polymeric light emitting diodes [4] and Schottky barrier devices [5]. These devices are made from thin polymer films, obtained by differnt techniques in different thicknesses.

Lithographic patterns of conducting polymers have been obtained by an number of different approaches, among them photoexposure of prepolymers followed by a wet chemical developing step [6], by wet photochemical or photoelectrochemical polymerization [7,8,9], by selective (dry) reactive ion etching in a microwave plasma in the presence of a suitable etchant [10], and by photochemically induced doping [11].

The availability of powerful lasers in the UV region (excimer laser), in the visible (ion lasers), and in the near infrared (solid-state lasers), has made it attractive and technically feasible to process films of electrically conducting polymers into lithographic patterns by different mechansims [12-16]. In the following, light- and laser-induced methods for lithographic patterning of conducting polymers will be discussed.

GENERATION OF POSITIVE PATTERNS

Electrically conducting polymers can be obtained via oxidative polymerization of suitable monomers, among them aromatic or heterocyclic systems [17], and electrically conducting composites may be obtained from conventional technical polymers by incorporation of electrically conducting polymers in the form of an interpenetrating network [18].
Among the oxidants, which are known to convert pyrrole into electrically conducting polypyrrole, there are some which are photosensitive to either visible or UV light or to other actinic radiation [19]. This makes it possible to expose a precomposite, consisting of a technical matrix polymer as a binder and the appropriate photosensitive oxidant, to radiation of the corresponding wavelength, upon which in the exposed areas the oxidant becomes selectively destroyed, whereas in the unexposed areas it remains unchanged. A typical example for such behavior are Fe(III)-salts, which are sensitive to visible or UV light, and which are being used as actinometers in photochemistry [20].

Potassium ferrioxalate, as one typical example, can be used as an actinometer from about 580 nm to well into the UV region of the spectrum. The primary photoreaction of this system proceeds with a quantum yield near unity [21] according to Equation 1:

$$Fe(III)(C_2O_4)_3 \xrightarrow{light} Fe(II)(C_2O_4)_2 + 2\ CO_2 \quad (1)$$

An additional advantage of this system is that neither the Fe(II)-ion nor its oxalate complex absorb significantly at the wavelength of the photoexposure. Consequently the photoactive oxidant bleaches upon exposure, thereby allowing a relatively high optical density of the photosensitive composite even when applying relatively thick coatings. This oxidant is soluble in water, and can be incorporated into water soluble matrix polymers (i.e. polyvinyl acetate) to yield photosensitive precomposites.

Using Fe(III)-chloride as the oxidant offers similar possibilities. Its good solubility in a variety of organic solvents makes it a suitable candidate to formulate precomposites together with water insoluble matrix polymers, such as polyvinylchloride (PVC) or polycarbonate (PC). The components of different successful systems are listed in Table I. For example, a solution consisting of PVC and FeCl$_3$ in tetrahydrofuran as the solvent was applied to sheets of glass via spin coating to yield films of a photosensitive polymer precomposite with thicknesses of a

Table I : COMPONENTS OF THE INVESTIGATED SYSTEMS

MATRIX POLYMER	OXIDANT	MONOMER
PVC	FeCl$_3$	pyrrole
PVA	NH$_4$Ce(NO$_3$)$_5$	N-methyl-pyrrole
PC	FeBr$_3$	N-phenyl-pyrrole
	Fe(NO$_3$)$_3$	thiophene
	Fe(ClO$_4$)$_3$	

few micrometers. Exposure of the dry films in the near UV through a mask converts the FeCl₃ in the exposed areas into FeCl₂, which has no potential to oxidize or polymerize monomeric pyrrole. Therefore, the latent image resulting from the photoexposure, can be developed into a "black and white" image, using pyrrole vapor. Thereby the remaining Fe(III)-salt in the unexposed areas induces the polymerization of the pyrrole monomer to an interpenetrating polymeric network.

No such polymerization reaction occurs in the exposed areas. Consequently, these photosensitive composites are positive working systems for polymer patterning. The polypyrrole renders the unexposed areas electrically conducting, whereby the electrical conductivity can be adjusted within a certain range via the concentration of the oxidant in the matrix polymer, i. e. via the FeCl₃ concentration. The scheme of the patterning procedure is shown in Figure 1. The two- dimensional patterns of conducting and insulating regions can be converted into three-dimensional structures either by reinforcing them through electroplating copper from aqueous solutions [22] or by removing predominantly the electrically conducting regions via reactive ion etching in an oxygen plasma [10].

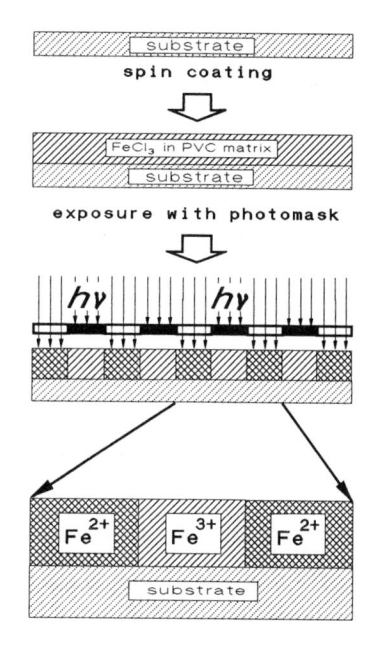

Figure 1: Scheme of patterning

GENERATION OF NEGATIVE PATTERNS

As mentioned above, common polymers can be rendered photosensitive using suitable sensitizers. Recently we found that chlorine containing polymers, in particular poly(vinylchloride), poly(chloroprene), poly(chlorostyrene) and poly(chloroacrylonitrile) become photosensitive to UV radiation in the range of 200 to 400 nm when exposing them to a vapor of heterocyclic monomers. For this purpose the polymers were dissolved in tetrahydrofuran, and spin coated onto sheets of glass or quartz in layers of 2 μm thickness. After thermal or vacuum enhanced drying, the samples were transferred into atmospheres of pyrrole or thiophene vapor for about ten minutes.

Another way of obtaining materials of comparable light sensitivity is to introduce solid monomers like dodecyloxythiophene into the polymer via a two component solution. This procedure yields clear polymer layers without considerable absorption in the visible, but with a strong absorption band below 350 nm. Table II summarizes the investigated systems. The light sensitive materials can be exposed to the radiation of a high pressure mercury lamp or of an excimer laser (351 nm XeF or 248 nm KrF). It was found that the period between removal of the samples from the monomer vapor and subsequent irradiation can be up to 24 hours in normal air without significant loss of sensitivity.

Irradiation of the light sensitive precomposite gives a black-brown material with an increased absorption in the visible and an absorption peak at 500 nm. This feature provides for a convenient check of the long term stability of the remaining light sensitivity of the pyrrole

Table II : COMPONENTS OF THE INVESTIGATED BINARY SYSTEMS

MATRIX POLYMER	MONOMER
polyvinylchloride	pyrrole
polychloroprene	thiophene
polychloroacrylonitrile	dodecyloxythiophene
polychlorostyrene	

system. The sensitivity of the polymer/monomer composites depends on the type of polymer. The exposed black-brown areas exhibit a polymer composite consisting of the starting matrix polymer interpenetrated by a polypyrrole or polythiophene network. The exposed and unexposed parts of the imaged polymer pattern have very different physical properties which permit their transformation into three dimensional lithographic patterns via different wet or dry procedures. Using tetrahydrofuran or a mixture of acetone/isopropanol/water, the latent image can be developed into a three dimensional resist structure owing to the different solubility of the exposed versus the unexposed regions. Figure 2 shows a corresponding microphotograph. A thin poly(chloroacrylonitrile) layer (thickness 2 µm) was put into a pyrrole vapor atmosphere for 5 hours. Afterwards the sensitized polymer layer was irradiated for 3 minutes using a 1000 W Hannovia high pressure mercury-xenon lamp and a chromium mask on quartz. The mask was put into close contact with the polymer layer. Tetrahydrofuran was used as the developer.

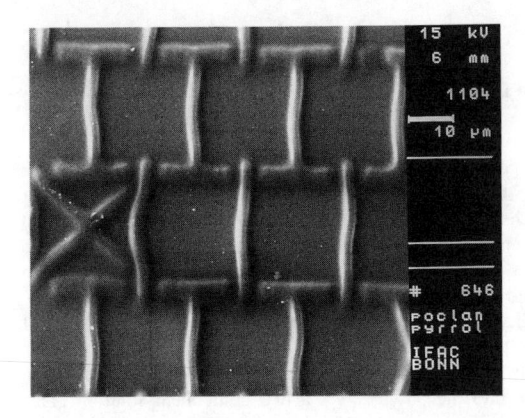

Figure 2 :
Electron micrograph of a resist pattern in poly(chloroacrylonitrile), developed with tetrahydrofuran.

On the other hand the exposed and unexposed regions have different ablation rates for excimer laser radiation. Therefore, the UV-exposed polymer composite can be processed subsequently by excimer laser radiation (248 nm KrF) to remove the unexposed polymer by ablation. In a third procedure the latent image of the polymer precomposite can be dry etched in an oxygen plasma reactor to remove the unexposed polymer regions, since the different regions (exposed/unexposed) show different etching rates. In a post development baking procedure at 800°C the polymer pattern can be converted into carbon for even more processing options. Figure 3 outlines the different methods for the manufacturing of three dimensional lithographic patterns.

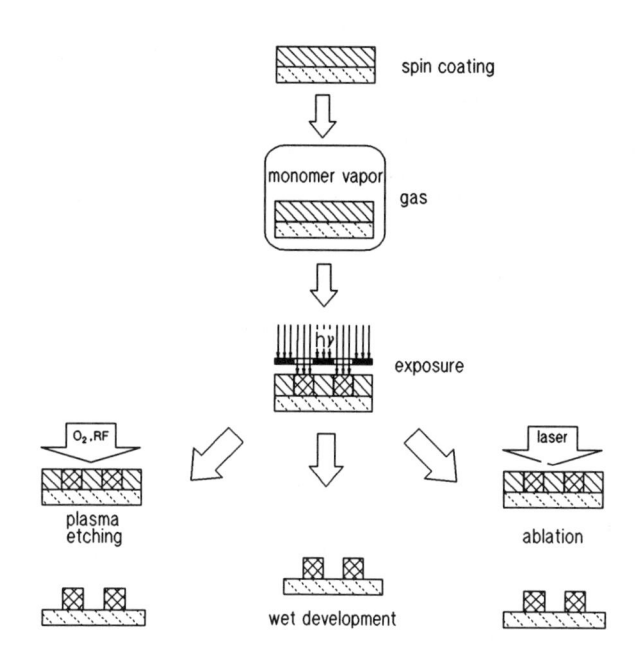

Figure 3 : Sequence of processing steps

LASER INDUCED PATTERNING OF POLY(BIS-ETHYLTHIO-ACETYLENE)

Lasers provide intense light and/or heat which can be focused onto a very small spot of a target. Combining the laser with a computer controlled positioning system yields a setup for polymer patterning [16]. Usually the polymer patterns obtained from photopolymers are electrically insulating, and they are typically used as resists for patterning metals or semiconductors in wet or dry processes after exposure. Recently it has been shown that Ar^+-laser irradiation of different polymers leads to highly conducting materials [13,14,15,24]. Whereas polyimide and chlorinated polyvinylchloride undergo graphitization processes, poly(bis-ethylthio-acetylene) (PATAC) seems to react via a photochemically induced cleavage of its sulphur containing sidegroups [15,25]. This mechanism opens up the possibility of generating an electrically conducting pattern without further processing, by irradiating thin layers of the insulating polymer PATAC with the focused beam of a 488 nm Ar^+-laser or a 351 nm XeF-excimer laser beam. Whereby the insulating PATAC becomes converted into a conducting form in the irradiated regions only.

The preparation and characterization of the PATAC has been described elsewhere [26]. Sheets of glass were plated with thin layers of PATAC by spin coating using a solution of PATAC in $CHCl_3$. The irradiation of the polymer samples was performed by a computer controlled linear positioning system coupled with a 2W-Ar^+-laser (488 nm). The irradiation equipment permits one to write a conducting pattern into the polymer substrate. The pattern itself was designed by a suitable layout editor. In the case of patterning the polymer by ablation with a XeF-excimer laser (351 nm), a stainless steel mask was used for imaging. The converted tracks were contacted with copper electrodes or suface mounted devices (SMD) using a conducting adhesive.

Exposure of thin layers of PATAC to 488 nm Ar^+-laser or 351 nm excimer-laser radiation converts the originally insulating material into an electrically conducting form. The conductivity of the polymer depends on the laser power and the scan velocity. In all cases the PATAC was converted in normal atmosphere.

The laser induced patterning of PATAC opens up new ways for manufacturing passive electronic components on a printed circuit board. Via this approach, we realized capacitors by converting the polymer into comb like structures in their conducting form. The width of a single track, as well as the distance between two tracks, was less than 20 µm. Using unconverted PATAC as the dielectric spacer material, we got capacitors with a specific capacity of more than 20 pF per cm^2. In a comparable procedure, resistors can be obtained by writing the tracks in a meandering form and tuning the irradiation conditions to result in a lower conductivity of the laser converted PATAC. These passive devices can be mated conveniently with SMD.

ACKNOWLEDGEMENT
This work has been supported in part by the Fonds der Chemischen Industrie.

REFERENCES

1. F. Garnier, G. Horowitz, X. Peng, D. Fichou, Adv. Mater. 2, 592 (1990)
2. G. Horowitz, X. Peng, D. Fichou, F. Garnier, J. Appl. Phys. 67, 528 (1990)
3. J. Paloheimo et al. Appl. Phys. Lett. 56, 1157 (1990)
4. J.H. Burroughes, D.D.C. Bradley, A.R. Brown, R.N.Marks, R.H. Friend, P.L.Burn, A.B. Holmes, Nature 347, 539 (1990)
5. R. Gupta, S.C.K.Misra, B.D.Malhotra, N.N.Beladakere, Appl.Phys.Lett.58, 51 (1991)
6. M.S.A.Abdou, G.A.Diaz-Guijada, M.I.Aroyo, S.Holdcroft, Chem.Mater. 3, 1003 (1991)
7. M. Okano, K. Itoh, A. Fujishima, K. Honda, J. Electrochem. Soc. 134, 837 (1987)
8. H. Segawa, T. Shimidzu, K. Honda, J. Chem. Soc., Chem. Commun. 1989 132
9. K. Yoshino et al. Jpn. J.Appl. Phys. 29, 1716 (1990)
10. J. Bargon, R. Baumann, P. Boeker, SPIE, (submitted)
11. B.M. Novak, E. Hagen, A. Viswanathan, L. Magde, Polym. Prepr. 31, 482 (1990)
12. C. Decker, J. Polym. Sci., C Polym. Lett. 25, 5 (1987)
13. J. Davenas, G. Boiteux, E.H. Adem, B. Sillion, Synth. Met. 35, 195 (1990)
14. H.K.Roth, H.Gruber, E.Fanghänel, A.Richter, W.Hörig, Synth. Met. 37, 151 (1990)
15. H.K. Roth, R. Baumann, M. Schrödner, H. Gruber, Synth. Met. 41, 141 (1991)
16. R. Baumann, J. Bargon, H.K. Roth, SPIE Vol. 1463, 638 (1991)
17. R.J. Waltman, J. Bargon, Can. J. Chem. 64, 76 (1986)
18. M.A. De Paoli, R.J. Waltman, A.F. Diaz, J. Bargon, J. Chem. Soc, Chem. Comm. 1984, 817; J. Polym. Sci., Chem. Ed. 23, 1687 (1985)
19. T. Ueno, H.D. Arntz, S. Flesch, J. Bargon, J. Macromol. Sci., Chem. A25, 1557 (1988)
20. N.J.Turro, Molecular Photochemistry, (W.A. Benjamin, New York 1965), 250
21. C.G. Hatchard , C.A. Parker, Proc. Royal Soc. A235, 518 (1956)
22. J. Bargon, T. Weidenbrück, T. Ueno, SPIE Vol.1262, 564 (1990)
23. C. Decker, Macromol. Chem., Macromol. Symp. 24, 253 (1989)
24. R. Baumann, J. Bargon, H.K. Roth, Mol. Cryst. & Liq. Cryst., (submitted)
25. A. Richter, J.M. Richter, N. Beye, E. Fanghänel, J. Prakt. Chem., 329, 811 (1987)

CHARACTERIZATION OF THERMOTROPIC LIQUID CRYSTALLINE POLYMER BLENDS BY POSITRON ANNIHILATION LIFETIME SPECTROSCOPY

Robert A. Naslund and Phillip L. Jones
Duke University, Department of Mechanical Engineering and Materials Science, Durham, North Carolina 27706
Andrew Crowson
U.S. Army Research Office, Research Triangle Park, North Carolina 27709-2211

ABSTRACT

Positron annihilation lifetime spectroscopy (PALS) is used to investigate polymer-based molecular composites of two separate thermotropic liquid crystalline polymers (TLCPs) and the thermoplastic matrix polyetherimide. Trends in PALS spectral components obtained isothermally at 20°C and 120°C as a function of TLCP content are discussed as a function of free volume and molecular structure. Differential scanning calorimetry was employed as a complementary technique to investigate the differences in miscibility of the two systems and possible effects on positron lifetime values.

INTRODUCTION

The increasing demand for lightweight, high performance materials has stimulated considerable research and development of reinforced polymer systems. In comparison to creating new polymers, blending polymers has emerged as a cost effective method of generating desired technological properties, especially for engineering purposes. Some primary areas of interest to this work include:

- Aromatic copolyesters, stiff "rod-like" molecules which exhibit nematic liquid crystalline behavior in the melt and become highly oriented in extended chain structures upon solidification.
- Polymer composites which exploit the high strength and modulus of aromatic copolyesters.
- Injection molded naphthalene moiety-containing thermotropic polyesters which have developed microfibril diameters on the order of 50 nm and fibril diameters on the order of 500 nm [1].

These systems offer a significant advantage over conventional composites as the difficulties associated with increased viscosity and fiber breakage are avoided in a thermoplastic/TLCP composite since the reinforcement fiber is liquid-like during processing and is formed *in situ* [2]. While these new materials offer significant advantages in physical properties, currently the cost of TLCPs is relatively high, this makes it essential to optimize the constituent compositions and processing conditions to yield the desired properties in TLCP blends.

The importance of free volume in the determination of viscoelastic properties has been the subject of considerable work in the field of polymer science. An understanding of these materials properties and the relationship between macroscopic mechanical properties and atomic scale free volume sites is necessary in any complete materials characterization of TLCP composites. The purpose of this study is to utilize PALS techniques to characterize the free volume state of these materials. PALS is a unique *in situ* analytical technique due to its internal probing nature and minimal perturbation between the probe and the system under study [3]. In molecular solids, it is well known that PALS is sensitive to the increase in the molecular mobility and the relaxation rate of the local motion of chains in response to an increase in temperature or a decrease in pressure [4, 5, 6, 7]. In addition, PALS has been demonstrated to be particularly sensitive to characterizing free volume in polymer materials.

An injected positron can enter into combination with an environmental electron to form a bound state known as positronium (Ps), which can best be thought of as an isotope of hydrogen where the nucleus has been replaced by a positron. Ps can exist in two states: para-positronium (p-Ps) and ortho-positronium (o-Ps). A free volume theory for the behavior of positrons in molecular solids has been introduced by Brandt et al. and states that the rate of positron or positronium annihilation is a function of the effective free volume of the solid [8]. Annihilation rate is dependent on the overlap of the positron component of the Ps atom with the lattice wave function. The larger the cavity in which the Ps atom is localized, the smaller is the overlap and the longer the positron will live before annihilation. Annihilation does not occur between the electron-positron pair forming the Ps atom, but rather occurs between the positron and another "bulk" environmental electron. Thus the bound electron in the Ps atom serves only to neutralize the charge of the positron. This constraint severely limits the Ps interaction with other bulk electrons and results in the increased lifetime that characterizes Ps annihilations. In polymers, PALS spectra are usually modeled using a three component decreasing exponential fit. The short lived component, τ_1 & I_1, is attributed to the annihilation of p-Ps, which shows no variation with temperature and structure, and possesses a lifetime of ~0.125 nsec. The intermediate-lived component, τ_2 & I_2, is attributed to the direct annihilation of positrons. This component is often referred to as "free" annihilation or "bulk" annihilation, and it possesses a lifetime in the 0.4 nsec range. The long-lived component, τ_3 & I_3, is attributed to the annihilation of o-Ps in free volume holes and possesses a lifetime between 0.5 to 5.0 nsec. It is the o-Ps which provides detailed information for the characterization of polymeric materials. Since the lifetime component is the reciprocal of annihilation rate, its magnitude increases with free volume expansion and is interpreted as being proportional to the size of the average free volume site. The corresponding intensity, I_3, is a measure of the fraction of annihilation events associated with τ_3 and is thus related to the amount of free volume.

EXPERIMENTAL

Samples used in this research were prepared by a novel injection molding technique developed at Virginia Polytechnic Institute and State University by Professor Donald G. Baird. The matrix material for all composites was Ultem 1000, a polyetherimide sold by General Electric Company. The TLCPs consisted of Vectra A-900, an aromatic copolyester based on 73% hydroxybenzoic acid and 27% 2-hydroxy 6-naphthoic acid, or HX4000, a polymer based on terephthalic acid, hydroquinone, and phenylhydroquinone.

Lifetime measurements were performed on two digitally stabilized EG & G Ortec fast-fast coincidence systems. The timing resolution of the systems were 280 and 300 psec, as determined using a prompt curve of a ^{60}Co source with energy windows of the discriminators set for ^{22}Na events. Each spectrum was modelled as the sum of three decaying exponentials with the 125 psec lifetime characteristic of the para-positronium self-annihilation fixed and a source correction. Source corrections were determined using annealed aluminum.

All positron samples were initially analyzed isothermally at 20°C. An average of five spectra, each consisting of approximately 10^6 counts, were analyzed by the computer modeling program PFPOSFIT [9]. The Vectra blend positron samples were then heated *in situ* at a rate of approximately 1.5°C/min to a final temperature of 120°C. An additional five lifetime spectra were collected isothermally at 120°C. The sample was then cooled at a rate of 1.25°C/min to a final temperature of 20°C. Three lifetime spectra were collected isothermally at 20°C and averaged. The difference in resolution of the two systems used in this study influences the absolute values obtained for lifetimes and intensities, but does not affect the trends noted.

The differential scanning calorimeter used in all experiments was a Seiko Instruments model 220C equipped with controlled liquid nitrogen cooling capabilities. Data analysis was performed using the Seiko version 2.4 computational software provided with the system. Sample weights ranged from approximately 5 to 10 mg. The thermal treatment employed for all samples consisted of heating at a rate of 20°C/min from 30°C to 380°C and holding for 20 minutes. The samples were subsequently cooled (quenched) at a rate of 50°C/min from 380°C to 30°C and held for 20 minutes; the samples were then reheated at 20°C/min from 30°C to 380°C, and finally were recooled to 30°C to reset the equipment for the next measurement. Data was analyzed on the second heat-up to eliminate the influence of any previous thermal histories due to processing [10]. All DSC experiments were carried out in air with no inert gas flowing through the sample chamber.

RESULTS

The o-Ps component for the HX4000 composites at 20°C is shown in Figure 1. The intensity data exhibit excellent agreement with a linearly increasing rule of mixtures (ROM) approximation. A linear regression of the experimental data yields a correlation coefficient of 0.997. The o-Ps lifetime values appear to be dominated by the TLCP phase with all blends possessing an average lifetime of 1.681 nsec with a standard deviation of only 5.7 psec. It is hypothesized that the free volume cavity size associated with the HX4000 TLCP remains constant for all blends and that this is the dominant site for o-Ps annihilation events. The data indicate that the pure HX4000, with a lifetime of 1.685 nsec, has a larger free volume cavity size than the pure Ultem with a lifetime of 1.558 nsec, a difference of 127 psec. A linear relationship between o-Ps intensity and TLCP content would be expected if the free volume cavity size is constant for all HX4000 blends, assuming that a certain number of free volume sites are associated with each weight percent increase in TLCP content.

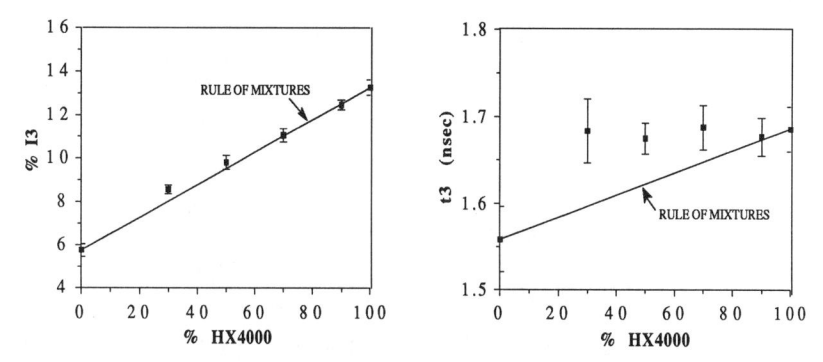

Fig. 1 Ortho-positronium in HX4000 blends at 20°C.

The o-Ps component for the Vectra composites at 20°C is presented in Figure 2. The intensity data exhibit a negative deviation from ROM behavior. In contrast to the above-described lifetime characteristics of the HX4000 blends, the Vectra blends do not show a comparable domination of the TLCP phase. In fact, the lifetime values seem to crudely approximate the ROM behavior. It should be noted that the 50 and 70 percent blends are relatively close at 1.534 and 1.485 nsec respectively and are more heavily influenced by the TLCP. Also, the 30% blend definitely shows the Ultem influence with a lifetime of 1.737 nsec. In contrast to the HX4000 system, where the lifetime of the TLCP is longer than the matrix, the lifetime of the Vectra is actually shorter than that of the Ultem.

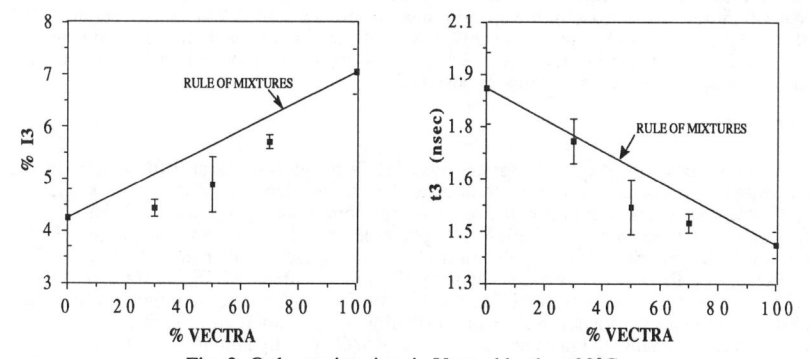

Fig. 2 Ortho-positronium in Vectra blends at 20°C.

In light of the domination of the lifetime values by the liquid crystalline phase in the HX4000 blends, the Vectra system was analyzed at 120°C with the hope of increasing the free volume capacity of the TLCP phase through a concomitant increase in temperature. The o-Ps component for the Vectra system at 120°C is presented in Figure 3. In contrast to 20°C, the data now more closely approximate the ROM relation with a definite negative deviation at 20°C shifting to a slightly positive deviation at 120°C. A more profound effect of temperature on the annihilation characteristics can be seen in the o-Ps lifetime component. The data exhibit a domination of the TLCP phase, with a trend similar to that shown in the HX4000 system at 20°C.

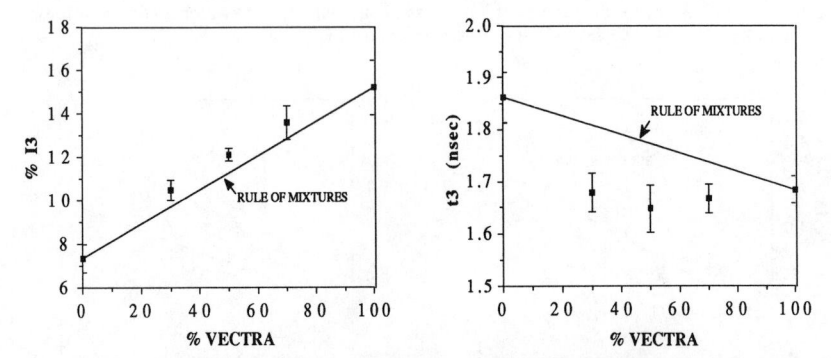

Fig. 3 Ortho-positronium in Vectra blends at 120°C.

The final set of PALS experiments was performed on the Vectra blends recooled from 120°C to 20°C. The lifetime parameters exhibited reasonable agreement with the data generated initially at 20°C with some degree of retained o-Ps intensity (free volume).

Differential scanning calorimetry was employed in this research as a complementary characterization technique to PALS. A linear decrease in the glass transition is exhibited with increasing concentration of TLCP for both blend systems. The decrease in the glass transition temperature for the 90% HX4000 blend was approximately 18.7°C. The decrease in the glass transition for the 70% Vectra blend was 4.5°C as compared with a decrease of 11.2°C for the 70% HX4000 blend. In addition to the lowering of glass transition temperatures, a broadening of the width of the glass transition measured calorimetrically was observed in both blend systems. The width of the glass transition of the Ultem matrix increased from 7.2°C for pure Ultem to 12.2°C for the 70% Vectra blend and to 11.5°C for the 70% HX4000 blend. The difference in the specific heat of the $C_{p_{glass}}$ and $C_{p_{liquid}}$ regions associated with the glass transition exhibited a linearly decreasing trend with increasing concentration of TLCP. In the HX4000 system, a decrease in the specific heat difference from 0.244 mJ/mg deg to 0.052 mJ.mg deg for the pure Ultem and the 90% HX4000 blend is observed. In the Vectra system, a decrease in the specific heat difference from 0.244 mJ/mg deg to 0.109 mJ.mg deg for the pure Ultem and the 70% Vectra blend is observed.

DISCUSSION

A comparison of PALS data for the two TLCP blend systems at 20°C results in the following observations. In striking contrast to the lifetime behavior of the HX4000 system, the Vectra system does not exhibit the constant free volume cavity size for all blends equal to that of the TLCP component. The HX4000 possesses an o-Ps lifetime value 127 psec larger than the Ultem matrix, while the Vectra lifetime is 479 psec smaller than the Ultem. The difference in o-Ps intensity between the pure Vectra and Ultem is 2.798%. In contrast, the difference between HX4000 and Ultem is 7.499%, indicating that the HX4000 has substantially more sites available for the formation and annihilation of o-Ps relative to the matrix. It is reasonable to infer that the o-Ps atoms will not only preferentially localize in the larger free volume sites, due primarily to their lower electron density, but will also preferentially localize in the dominant free volume site based on probability. It is hypothesized that the difference in o-Ps intensities, i.e. number of sites, is controlling lifetime characteristics.

The 30, 50, 70, and 100 percent Vectra blends at 120°C possess an average o-Ps lifetime of 1.670 nsec with a standard deviation of only 16 psec. The data is now TLCP-dominated in much the same way as the HX4000 systems at 20°C. A comparison of the data between the Vectra systems at 20°C and 120°C reveals that the Ultem intensity increased by 3.081%, the Ultem lifetime decreased by 36 psec, the Vectra intensity increased by 8.123%, and the Vectra lifetime increased by 265 psec. Essentially, the lifetime of the Ultem matrix remains constant whereas there is a substantial increase in the lifetime of the TLCP as a consequence of the increase in temperature.

The change in annihilation characteristics with temperature for the Ultem matrix can be explained in terms of its glass transition and the o-Ps lifetime temperature dependence. There exists a temperature below which all motions, including molecules and chains, are completely frozen and no detectable variance of free volume with temperature is observed by positron annihilation lifetime spectroscopy [3, 4]. This point has been termed the sub-glass transition temperature, T_s. Above T_s two distinct variations of τ_3 with temperature are observed. A weaker variation is exhibited below T_g with a stronger variation occurring above T_g, with the most rapid change in τ_3 occurring at T_g. The region between T_s and T_g has been attributed to side chain motion mainly confined to the single chain segment and to the increase in segmental mobility of the macromolecular chains within the amorphous region of a semicrystalline polymer which are not constrained by the crystalline phase [3, 4]. At the glass transition temperature, the molecular motion reaches a point where the inclusion of main-chain segmental motion is possible [3]. Above T_g, the matrix becomes rubbery and the fractional free volume content increases dramatically with temperature. In the region above T_g, the trapped positronium experiences an environment where the electron density is rapidly decreasing with increasing temperature. Furthermore, the region is characterized by an increase in segmental mobility resulting from the diminishing constraint imposed by the crystalline phase upon the surrounding glassy regions. The experimentally determined glass transition was approximately 219°C for pure Ultem. At 120°C, 100°C below the glass transition, it is reasonable to expect that the molecular motion of the system has not yet increased to a level detectable by positron annihilation lifetime spectroscopy. It is believed that the Ultem lifetime behavior exhibited lies below the sub-glass transition temperature, T_s, for all temperatures between 20°C and 120°C. Accordingly, the lifetime of the o-Ps component will not change. This conclusion is supported by the data exhibited in figures 2 and 3.

In contrast to the Ultem matrix at 120°C, the liquid crystalline polymer, Vectra, shows a dramatic increase in both lifetime and intensity values. The glass transition of Vectra, determined experimentally, is approximately 150°C. Therefore, the temperature of data generation (120°C) is only 30°C below the glass transition of Vectra. It is reasonable to expect that the molecular motion of the TLCP has increased measurably. It is hypothesized that at 120°C, the lifetime value can be found somewhere on the weaker variation of τ_3 with temperature between T_s and T_g.

The Vectra lifetime data at 120°C show a difference in intensity between the TLCP and matrix of 7.84%. In contrast, the lifetime data exhibit only a 178 psec difference. It is hypothesized that the increase in intensity for the Vectra component, as a result of the 100°C increase in temperature, is responsible for the dramatic change in the lifetime data component. As in the HX4000 system at 20°C, the TLCP dominates the number of free volume cavity sites. Although the matrix has a larger cavity size by 178 psec, the difference in intensity points to the preferential annihilations of o-Ps in Vectra free volume sites. The probability of a diffusing positronium atom's finding a Vectra free volume site is much greater than that of finding an Ultem site.

The lifetime value for the recooled Ultem is only 2 psec less than that of the initial Ultem at 1.898 nsec. The lifetime value for the recooled Vectra is 22 psec less than that of the initial Vectra at 1.419 nsec. The 30, 50, and 70 percent Vectra blends had an average of 70.7 psec of retained lifetime as compared with the initial data at 20°C. A comparison of the recooled o-Ps intensity values to the initial values at 20°C reveal 1.366% retained intensity for the pure Ultem and 0.297% retained intensity for the pure Vectra. The 30, 50, and 70 percent Vectra blends possessed a retained intensity value of 1.205%, 1.243%, and 0.254% respectively. The above data seem to indicate that the free volume cavity size, τ_3, is recovered almost instantaneously as expressed in the rather small differences in lifetime values. On the other hand, the intensity, or the number of free volume cavity sites, seems to recover more quickly for the pure Vectra and the 70% composite than for the pure Ultem and the lower weight percent blends as expressed in the retained intensity data. These data seem to indicate that the liquid crystalline polymer relaxes more quickly than the matrix.

A qualitative analysis of the mean free volume cavity sizes of the various polymer examined yields the following relationship:

$$\tau_3(\text{HX4000}) > \tau_3(\text{Ultem}) > \tau_3(\text{Vectra}).$$

The above relation can be explained in terms of the stereochemistry of the molecules. The intermolecular free volume associated with the hindrance in packing of the HX4000 molecules, as a consequence of the pendant phenyl groups, is hypothesized as being responsible for additional ortho-positronium localization and thus the large mean free volume exhibited. The Vectra TLCP, by contrast, is a highly linear molecule with no side groups. Thus, one would expect such a molecule to have a smaller fraction of free volume. Finally, the Ultem amorphous matrix would be expected to possess an intermediate free volume cavity size, largely due to the central carbon linking aromatic ether groups.

The determination of the degree of compatibility between blends of an amorphous polymer and a semi-crystalline polymer is often detected in the glass transition behavior of the two polymers. Complete compatibility is manifested in the appearance of a single glass transition whose temperature is intermediate between the glass transition of the two component polymers [11]. In contrast, blends of incompatible polymers are expected to segregate into distinct phases each exhibiting a glass transition identical in temperature and width to those of the unblended components [11]. An intermediate case exists, which is in fact the behavior exhibited by the liquid crystalline polymer blends in this research, where a partial mixing between the components results in a shift in one or both of the individual glass transitions [11]. In addition to the suppression of glass transition temperatures in compatible blends, a broadening of the width of the transition measured calorimetrically is often observed. Based on the decrease in glass transition temperature of the polyetherimide component and the broadening of the transition width, it is hypothesized that both blends exhibit partial miscibility with the HX4000 system possessing greater miscibility.

SUMMARY

Positron annihilation lifetime spectroscopy allows the examination of free volume cavity sizes and distribution in thermoplastic/TLCP composites on the submicron level. Variations in both spectral components is observed as a function of temperature and TLCP content. It has been shown that these differences between the matrix and reinforcing phase has substantial influence on the localization and annihilation of o-Ps. The more miscible HX4000 composites exhibited TLCP dominated lifetime characteristics at room temperature. DSC used as a complementary characterization technique to PALS has shown that the greater miscibility exhibited by the HX4000 system over the Vectra system provides the potential to form composites on a size scale closer to the molecular level.

ACKNOWLEDGMENTS

The authors wish to thank Dr. D.G. Baird of VPI&SU for providing the samples used in this research. The financial support of the Army Research Office under Contract No. DAAL03-89-C-0020 is also gratefully acknowledged.

REFERENCES

1. L.C. Sawyer and M. Jaffe, Journal of Materials Science 21, 1897-1913 (1986).
2. R.A. Weiss, W. Huh and L. Nicolais, Polymer Engineering and Science 27, 684-691 (1987).
3. Y.C. Jean, Microchemical Journal 42, 72-102 (1990).
4. J.H. Lind, P.L. Jones and G.W. Pearsall, Journal of Polymer Science: Part A: Polymer Chemistry Edition 24, 3033-3047 (1986).
5. A.D. Kasbekar, P.L. Jones and A. Crowson, Journal of Polymer Science: Part A: Polymer Chemistry 27, 1373-1382 (1989).
6. S.Y. Chung, S.J. Tao and T.T. Wang, Macromolecules 10, 713-715 (1977).
7. P. Kindl and H. Sormann, Phys. Status Solidi A 66, 627-633 (1981).
8. W. Brandt, S. Berko and W. Walker, Physical Review 120, 1289-1295 (1960).
9. W. Puff, Computer Physics Communications 30, 359-368 (1983).
10. A. Kohli, N. Chung and R.A. Weiss, Polymer Engineering and Science 29, 573-580 (1989).
11. W.J. MacKnight, F.E. Karasz and J.R. Fried, in Polymer Blends, edited by R. Paul and S. Newman (Academic Press, New York, 1978), pp. 185-242.

SMALL ANGLE NEUTRON SCATTERING STUDIES OF SINGLE PHASE IPNS

Barry J. Bauer[*], Robert M. Briber[*], Shawn Malone[**] and Claude Cohen[**]
[*] National Institute of Standards and Technology, Polymers Division, Gaithersburg, MD, 20899.
[**] School of Chemical Engineering, Cornell University, Ithaca, NY, 14853.

ABSTRACT

Interpenetrating polymer networks (IPNs) have been synthesized from polymers that form miscible polymer blends. Full, semi-I and semi-II IPNs made from polystyrene-d8 and poly(vinylmethylether) can be made to phase separate by incorporating low levels of crosslinking. However, blends of these polymers have a negative Flory-Huggins interaction parameter, making them highly miscible. This indicates that formation of IPNs favors phase separation relative to blends.

IPNs made from polystyrene-d8 and polystyrene-h8 show that increased crosslink density also destabilizes the mixture as shown by small angle neutron scattering.

IPNs have also been made by crosslinking end functionalized polydimethylsiloxanes in the presence of nonfunctionalized, deuterated siloxanes. These IPNs are also destabilized by increasing crosslink density, suggesting that the destabilization is due to the network, and not to the particular type of network forming reaction.

INTRODUCTION

Interpenetrating polymer networks (IPN) are synthesized from two different types of polymers. Sequential IPNs are made in two independent polymerizations, the second of which involves polymerization of monomers which are dissolved in the first polymer. Due to the considerable entropy of mixing of the small monomer molecules, there are many combinations of polymers and monomers that will dissolve in each other forming a single phase. Upon polymerization of the second monomer, entropy of mixing is lost and the two polymeric components usually phase separate, forming the two-phase morphology commonly seen in IPNs.

There are a few polymer pairs that form miscible blends, that is, even at high molecular weight, these polymers form a single phase mixture. IPNs made from polymers that form miscible blends can be used to study the effects of crosslinking on the stability of the IPNs. Small angle neutron scattering (SANS) has been used to measure the stability of blends and IPNs made from the same polymers.

THEORY

The Flory-Huggins interaction parameter χ for a miscible polymer blend can be measured by applying the random phase approximation [1] to SANS results. Equation 1 gives the relationship used with $S(q)$, the scattered intensity; k_n, the contrast factor; v_i, the volume of an i unit; N_i, the degree of polymerization of i; ϕ, the volume fraction of A; $S_i(q)$, the Debye function of polymer i, and v_0, the reference volume. The usual way of expressing the interaction parameter is as χ/v_0 which is independent of the size of the reference volume.

$$\frac{k_n}{S(q)} = \frac{1}{v_A N_A \phi S_A(q)} + \frac{1}{v_B N_B (1-\phi) S_B(q)} - \frac{2\chi}{v_0} \qquad (1)$$

When there are labeled chains in an unlabeled matrix (a network in this case) the total scattering function cannot be expressed since a simple Debye function cannot be used to describe the distribution of units in the network. The zero angle scattering can be expressed in terms of polymer variables as in equation 2.

$$\frac{k_n v_0}{S(0)} = \frac{B}{N_c \phi} + \frac{A \phi_s^{2/3}}{N_c \phi^{5/3}} + \frac{1}{N_B(1-\phi)} - 2\chi \qquad (2)$$

N_C is the degree of polymerization between crosslinks, ϕ_S is the volume fraction of network where the chains are relaxed, and it is usually assumed that the chains are relaxed when formed so that $\phi_S = \phi$ in a semi-II IPN. In a semi-I IPN the first polymerization is usually done in bulk so that $\phi_S = 1$. A and B are constants of the network, being 1 and 0.5 in this case [2].

EXPERIMENTAL

Semi-II IPNs of polystyrene-d8 (PSD) and poly(vinylmethyl ether) (PVME) were made by dissolving linear PVME in styrene-d8 containing divinyl benzene (DVB) and a free radical initiator and polymerizing thermally. The IPNs of PSD and standard polystyrene (PSH) were made in a similar manner. Details of the synthesis are given elsewhere [3-4].

Semi-I and full IPNs were synthesized by first crosslinking the PVME by exposure to gamma irradiation from a Co[60] source. The crosslink density was calculated by measuring the change in the molecular weight of a PVME sample exposed to doses low enough to cause branching but not high enough to cause crosslinking. Size exclusion chromatography (SEC) was then used to measure the loss of the original material due to branching. The rate of bond formation is used to calculate the crosslink density in PVME samples given a much higher dose. Semi-I IPNs are then formed by swelling the networks with styrene-d8 and initiator and polymerizing thermally as was done with the semi-II IPNs. Full IPNs were made by swelling the networks with styrene-d8 containing DVB and polymerizing.

Siloxane networks were prepared by mixing unfunctionalized linear chains, containing matched pairs of hydrogen containing and deuterium containing chains, with functionalized chains. The functionalized chains were then linked together forming tetrafunctional networks as has been described previously [5].

The SANS was performed at the NIST 8 meter facility. The wavelength of the incident neutron beam was 9 Å with a $\Delta\lambda/\lambda$ of 25% as determined by a rotating velocity selector. Scattering was done from 30°C to 150°C. Data were collected on a two-dimensional detector, corrected for empty cell and incoherent scattering, and placed on an absolute scale by use of a calibrated secondary standard.

RESULTS AND DISCUSSION

The scattering results for four PVME/PSD semi-II IPNs are shown in figure 1. Samples containing 0.0, 0.1 and 0.3 % DVB were taken at 100°C and 1.0% at 70°C. At these temperatures the χ parameter is negative, causing blends of linear polymers to be miscible. As the amount of crosslinking increases, the total scattering increases, showing that the blends become destabilized. The sample with 1.0% DVB shows scattering typical of phase separated samples [3]. Figure 2 gives similar results for PSH/PSD semi IPNs. With increased DVB content, the scattering increases until phase separation occurs [4,6].

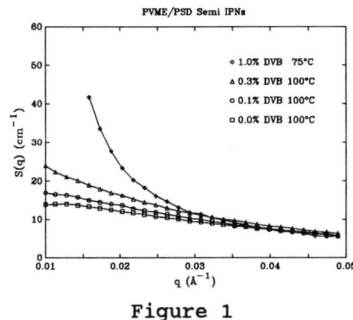

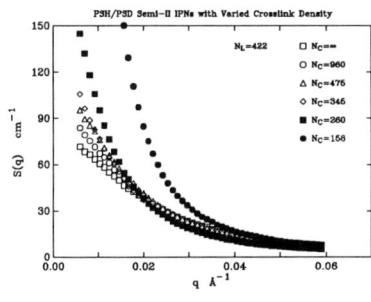

Figure 1

Figure 2

The results of figures 1 and 2 are typical of semi-II IPNs formed by second stage vinyl-divinyl copolymerizations. In all cases studied, polymers forming miscible blends could be made to phase separate by increasing the crosslink density. Semi-I and full IPNs have a network present when the second polymerization is taking place, and perhaps this network will affect the phase stability in a different way.

Figure 3 gives the SEC results of a PVME sample with a molecular weight Mw = 120,000 and a polydispersity Mw/Mn of 1.2. Also shown are samples exposed to various doses of gamma irradiation. As the dose rate increases, the main peak decreases and signal appears at lower elution volume (higher molecular weight). This is branched material formed by the joining of two original PVME molecules. The rate of loss of the original material can be used to calculate the number of bonds formed for a given dose by a procedure described elsewhere [7]. There is no increase in the signal to the right of the original peak indicating no buildup of low molecular weight material. This suggests that chain scission is not an important factor compared to chain branching. Figure 4 is a plot of the decrease in the intensity of the main peak with dose. It is linear over a wide range and gives the relationship $N_C = 92.55/D$ where D is the dose in MGy.

PVME samples were exposed to 0.4 MGy doses giving $N_C = 231$. Some networks were then swollen with styrene-d8 and polymerized forming semi-I IPNs and others were swollen with styrene-d8 containing DVB and polymerized forming full IPNs. SANS experiments were conducted on these samples at a variety of temperatures. Figure 5 shows the results of the semi-I sample plotted as $S(q)^{-1}$ versus q^2. There is a large amount of curvature in the plots suggesting that phase separation has occurred. Figure 6 replots the results as $S(q)^{-\frac{1}{2}}$ versus q^2. In this form the plots are much more linear, again suggesting that phase separation has occurred. The results for the full IPNs were qualitatively the same.

No semi-I or full IPNs were prepared that remained single

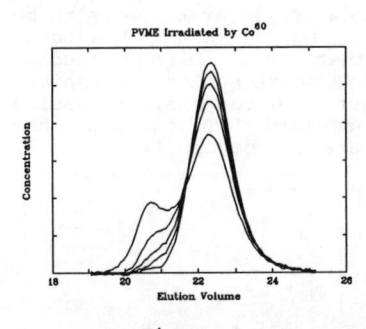

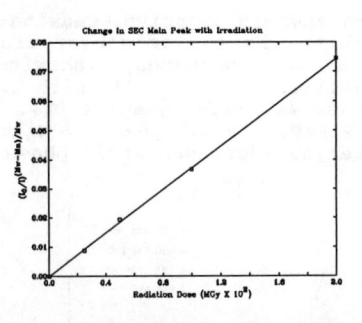

Figure 3

Figure 4

phase. It is possible that there may be ranges of compositions and crosslink densities that would produce single phase samples, but it appears that the crosslinking of the first component promotes phase separation, similar to the effect of crosslinking the second component.

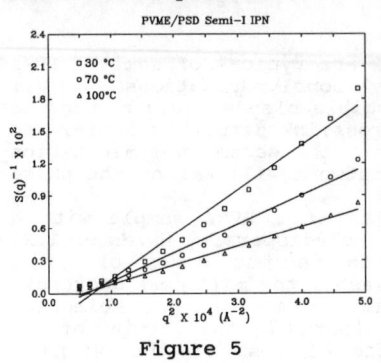

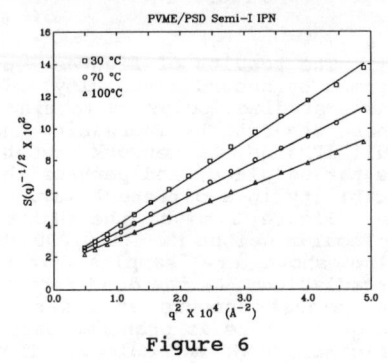

Figure 5

Figure 6

DVB copolymerization can form networks with a wide distribution of crosslink densities [8] and this may affect the stability. In an effort to make more uniform networks, end linked siloxane IPNs were made. Siloxane networks are made from siloxane chains carrying a functionality at each end that can react with a linking agent to form model networks. The prepolymers have narrow molecular weight distribution forming networks that are much more uniform than DVB produced ones. Other siloxane chains can be produced without the terminal functionalization, so that no grafting reaction is possible to link them to the network.

Figures 7 plots the scattering intensities of siloxane IPNs made from deuterated linear chains in networks of tetrafunctionally crosslinked protonated siloxane chains. Both samples had deuterated chains with M_n = 12800 and M_w = 17300 as determined by SEC. The E-1 sample had network chains with M_n = 53500 and contained 0.239 volume fraction of linear chains. The G-1 sample had network chains with M_n = 10900 and a volume fraction 0.242 of linear chains.

There is more scattering intensity from the G-1 sample,

which contained a higher crosslink density. This sample has been destabilized with respect to sample E-1, which has a much lower crosslink density. This result is consistent with the results shown in figures 1 and 2 for semi-II IPNs made from PVME/PSD and PSH/PSD respectively. In all three cases increasing the crosslink density destabilized the system, even though different polymer types and different crosslinking methods were used.

The zero angle scattering intensity can be used with equation 2 to calculate the interaction parameter necessary to account for the results. This is similar to the procedure used to account for the results of swelling measurements made by McKenna et al. [9] This calculated value of the interaction parameter represents a value required to make the theory described by equation 2 fit the data and is a convenient way of observing the deviation of the experiment from theory.

Figure 8 is a plot of χ/V_0 versus $1/N_C$ for semi-II IPNs of siloxanes, PVME/PSD, and PSH/PSD. The intercept at $1/N_C = 0$ is for blends of linear polymers and is very close to zero for blends of protonated and deuterated version of the same polymer. The PVME/PSD interaction parameter is known to be negative and sizable at 100°C [10]. All have similar slopes, indicating that the effect of N_C on the scattered intensities is the same.

Recent theory has incorporated heterogeneities of the network into the scattering equations [11-12]. While the increased scattering seen in these IPNs is predicted, it is surprising that the results for PSH/PSD and siloxane IPNs are so similar. The molecular weight distributions between crosslinks in the polystyrene networks is expected to be broad, certainly Mw/Mn > 2, while for the E series Mw/Mn = 1.27 and for the G series Mw/Mn = 1.19. It seems likely that the polystyrene networks are much more heterogeneous than the siloxane networks for this reason.

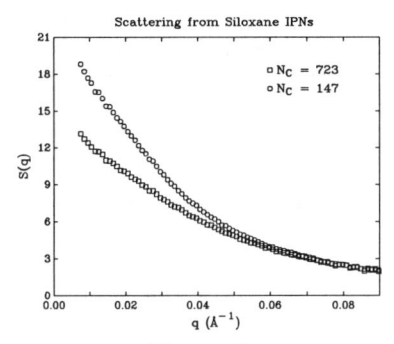

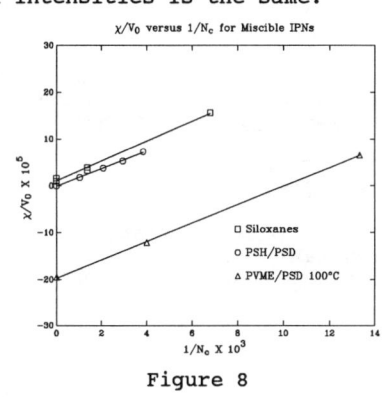

Figure 7

Figure 8

CONCLUSIONS

IPNs made from polymers that form compatible blends are destabilized compared to the blends and can be made to phase separate by increasing the crosslink density. This is the case when the crosslinking takes place in either the first or the second polymerization, that is, semi-I, semi-II, and full sequential IPNs are all destabilized.

The deviations in scattered intensity from that of theory appears to be the same in samples with networks made from DVB copolymerizations and networks made by end linking narrow molecular weight distribution chains.

REFERENCES

1. P. G. de Gennes, <u>Scaling Concepts of Polymer Physics</u>, Cornell University Press, New York, 1979, Ch. IV.

2. P. J. Flory, <u>Principles of Polymer Chemistry</u>, Cornell University Press, Ithaca, New York, 1953, Ch. XIII.

3. B. J. Bauer, R. M. Briber, and C. C. Han, Macromolecules, <u>22</u>, 940 (1989).

4. R. M. Briber and B. J. Bauer, Macromolecules, <u>24</u>, 1899 (1991).

5. M. R. Aven and C. Cohen, Makromol. Chem., <u>189</u>, 881 (1988).

6. R. M. Briber and B. J. Bauer, (Mater. Res. Soc. Proc. <u>171</u>, Boston, MA 1990) pp. 203-210.

7. R. M. Briber and B. J. Bauer, Macromolecules, <u>21</u>, 3296 (1988).

8. S. Candau, J. Bastide, and M. Delsanti, Adv. Polym. Sci., <u>44</u>, 27 (1982).

9. G. B. McKenna, K. M. Flynn, Y.-H. Chen, Polymer Communications, <u>29</u>, 272 (1988); Macromolecules, <u>22</u>, 4507 (1989); Polymer, <u>31</u>, 1937 (1990).

10. C. C. Han, B. J. Bauer, J. C. Clark, Y. Muroga, Y. Matushita, M. Okada, q. Tran-Cong, T. Chang, and I. C. Sanchez, Polymer, <u>29</u>, 2002 (1988).

11. J. Bastide, L. Liebler, and J. Prost, Macromolecules, <u>23</u>, 1821 (1990).

12. A. Onuki, <u>Formation, Dynamics and Statistics of Patterns</u>, K. Kawasaki ed., World Science publishers (1989); additional communications to be published.

PART II

Organic/Inorganic Composites

HIGH GLASS CONTENT NON-SHRINKING SOL-GEL COMPOSITES VIA SILICIC ACID ESTERS

MARK W. ELLSWORTH AND BRUCE M. NOVAK*
University of California, Berkeley, Dept. of Chemistry, Berkeley, CA 94720
and the Center for Advanced Materials, Lawrence Berkeley Laboratory, Berkeley, CA.

Abstract

The application of the sol-gel process for the formation of inorganic-organic composites has received a great deal of attention in recent years. We have focused our efforts in this area toward the development of simultaneous polymerization/sol-gel reactions for the developement of inorganic-organic composites with a wide range of polymers. Further research in this area has led to the design and synthesis of tetraalkoxysilanes, and more recently, poly(silicic acid esters) possessing polymerizable alkoxides for the synthesis of non-shrinking sol-gel composites.

Introduction

Although the sol-gel process was first demonstrated over a century ago, major applications of this technique towards the production of ceramic oxides have only been reported in the past few decades. Major improvements in the processing of sol-gel glasses have resulted in important applications such as high purity optical materials, fiber optics, and coatings.[1] We, as well as other researchers, have been interested in the application of the sol-gel process for the production of hybrid inorganic/organic composites.

The sol-gel process involves the sequential hydrolysis and polycondensation of silicon alkoxides (usually tetramethylorthosilicate (TMOS) or tetraethylorthosilicate (TEOS)) in acidic or basic aqueous solution with a suitable cosolvent to form a solvent swollen inorganic gel. After slow evaporation of the cosolvent, liberated alcohol, and excess water, a low density glass called a xerogel results (Scheme I). This can be taken as the final product, or the xerogel can be sintered at high temperatures to produce densified glass.[1] Because the initial sol-gel reaction occurs at ambient temperatures and pressures, a variety of organic molecules and polymers can be incorporated into the inorganic network to produce organically modified glass. Organic polymers such as poly(tetramethylene oxide),[2] poly(acrylates),[3] and poly(siloxanes)[4] have been employed in the synthesis of hybrid inorganic/organic composites via the sol-gel process.

Two major limitations in the synthesis of sol-gel composites are the difficulty of finding polymers that are soluble in the aqueous sol-gel solution and the ubiquitous shrinkage that results from the evaporation of excess solvents and water. In our research, we have attempted to address both of these issues by developing *in situ*, synchronous polymerization routes into these hybrid materials.[5] The thermodynamic instabilities associated with the mixing of two dissimilar phases were overcome kinetically by simultaneously forming the organic polymer and the inorganic network. This enabled us to produce composites with polymers that are normally incompatible with the sol-gel solutions by interpenetrating the two networks to such an extent that large scale phase separations do not occur. We have designed tetraalkoxysilanes, and more recently, poly(silicic acid) esters possessing polymerizable alkoxides which allow for the production of non-shrinking sol-gel composites. With these new approaches, inorganic/organic composites with a wide range of properties can now be synthesized.

Scheme I

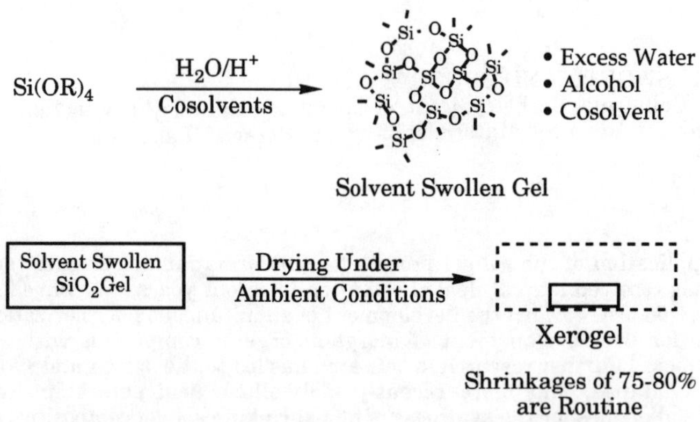

Si(OR)$_4$ $\xrightarrow[\text{Cosolvents}]{\text{H}_2\text{O/H}^+}$ Solvent Swollen Gel
- Excess Water
- Alcohol
- Cosolvent

Solvent Swollen SiO$_2$ Gel $\xrightarrow[\text{Ambient Conditions}]{\text{Drying Under}}$ Xerogel

Shrinkages of 75-80% are Routine

Results and Discussion

One important prerequisite in the production of inorganic/organic composites is that the organic polymer must be soluble in the aqueous sol-gel solution. Unfortunately, only a limited number of polymers are soluble in sol-gel solutions. This limitation became apparent to us in our own work with highly functionalized aqueous ring-opening metathesis polymers (ROMP polymers).[5] Although many of these polymers are readily soluble in organic solvents, the addition of water or acid to these solutions immediately results in precipitation of the polymer. Even in the case of water soluble polymers, large scale phase separation occurs during gelation, resulting in opaque, brittle materials. To overcome this problem, we investigated the formation of simultaneous interpenetrating networks (SIPN's), where both the organic polymer and inorganic network are formed at the same time (Scheme II).[5] This simultaneous route allows for the homogeneous incorporation of both soluble but incompatible polymers *and completely insoluble polymers* into the inorganic network.

Scheme II

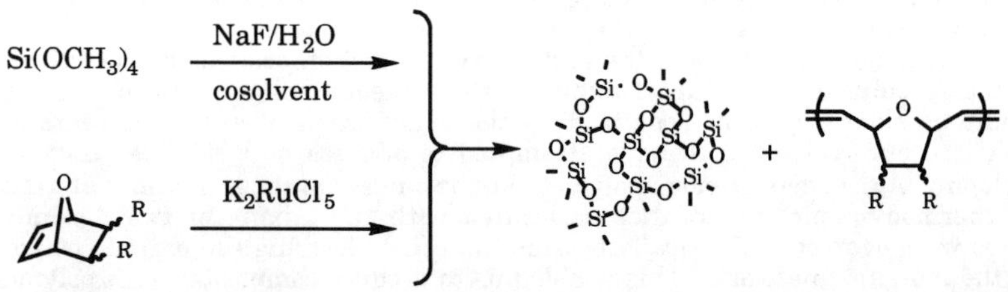

Si(OCH$_3$)$_4$ $\xrightarrow[\text{cosolvent}]{\text{NaF/H}_2\text{O}}$

$\xrightarrow{\text{K}_2\text{RuCl}_5}$

A significant drawback to the sol-gel process is the shrinkage that occurs during the evaporation of excess water, liberated alcohol, and cosolvent. Shrinkages of 75-80% are common, and as a result of this shrinkage, gels must be dried at extremely slow rates to prevent cracking and deformation of the xerogel. In an attempt to design non-shrinking sol-gel composites, we have synthesized tetraalkoxysilanes possessing polymerizable alkoxides from SiCl4 and various polymerizable alcohols (eq. 1).[6] By using the tetraalkoxysilane derivative, a polymerizable monomer as the cosolvent, and a stoichiometric quantity of water, all components in the reaction mixture contribute to the inorganic network or the organic polymer. Since the cosolvent and liberated alcohol are polymerized, drying is unnecessary and only minor gel shrinkage occurs (Scheme III). Since the composite synthesis is essentially a bulk polymerization, small shrinkages do occur. The important aspect of this work is that the large scale shrinkages associated with the drying step are now eliminated.

$$SiCl_4 + ROH \xrightarrow[THF]{Et_3N} Si(OR)_4 \quad (1)$$

R =

Since our first reports on the use of tetraalkoxysilanes possessing polymerizable alkoxides for the synthesis of non-shrinking sol-gel composites, we have focused our efforts towards increasing the glass content in these materials.[7] The stoichiometry in the tetraalkoxysilanes limits the maximum glass content in the non-shrinking composites to 10-18%. By increasing the number of Si-O-Si bonds, or branch points, in the sol-gel precursor, we can increase the glass content in the final composite. Poly(silicic acid) is a convenient choice for a precursor with the branching structure necessary to produce high glass content composites. Neglecting end group effects, the weight percent glass in composites derived from substituted poly(silicic acid) esters is given by equation 2, where MW is the molecular weight of the substituted alcohol; n, m, and p are the percent Q^2, Q^3, and Q^4 silicon centers respectively; and x and x' are the percent alkoxide substitution on the Q^2 and Q^3 silicon centers respectively.[8]

$$\% \text{ Glass} = \left[\frac{1}{n}\left(1 + x\left(\frac{2MW}{60.1}\right)\right)\right]^{-1} + \left[\frac{1}{m}\left(1 + x'\left(\frac{MW}{60.1}\right)\right)\right]^{-1} + P \quad (2)$$

Silicic acid is synthesized by the reaction of sodium metasilicate with water in acidic aqueous solution.[9] The molecular weights of poly(silicic acid) can be varied from 5,000 to 2,000,000 simply by increasing the reaction time (table 1). Unfortunately, the number of branch points (Q^4 species) does not change to any great degree with increased molecular weight. A more convenient and effective method for increasing branching is to adjust the acid concentration in the sodium metasilicate reaction (table 2). Branching ratios were determined by ^{29}Si NMR (figure 1).[10]

Scheme III

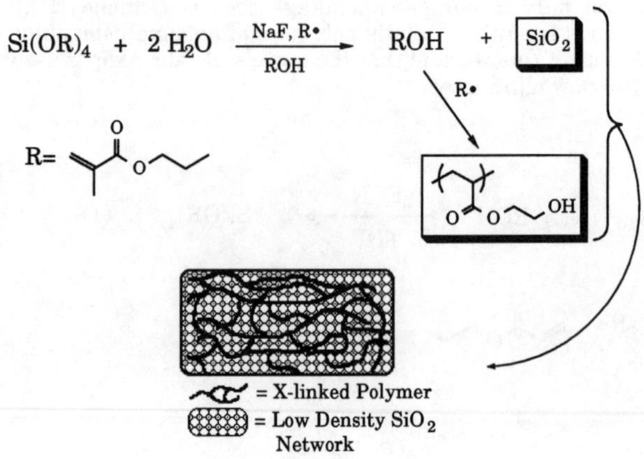

rxn. time	**M$_w$**	**M$_n$**	**PDI**	**yield**
1h	8138	5044	1.61	80-90%
24h	44,719	19,742	2.27	80-85%
48h	183,145	26,293	6.97	75-80%
72h	7,770,902	1,876,154	4.41	50-60%
+72h	-insoluble particles-			----

Table 1. Molecular weight data for benzyl silicic acid esters relative to polystyrene.

M$_w$	**[HCl]**	**Q^1**	**Q^2**	**Q^3**	**Q^4**
8130	3.0 M	0.7%	11.0%	49.3%	39.0%
44,719	3.0 M	<0.1%	10.2%	55.0%	34.8%
183,145	3.0 M	<0.1%	10.0%	50.0%	40.0%
7,770,902	3.0 M	<0.1%	11.0%	50.0%	50.0%
5000	6.0 M	<0.1%	4.8%	33.0%	62.2%

Table 2. Q values determined by ^{29}Si NMR for silicic acid oligomers.

The literature method for silicic acid esterification calls for azeotropic distillation of the alcohol used for the substitution.[9] In the cases of the thermally sensitive acrylates and synthetically valuable ROMP monomers, an alternative route was needed. Instead of the usual azeotropic distillation of the alcohol, the water generated in the esterification was removed by azeotropic distillation of tetrahydrofuran (THF) (Scheme IV). This procedure elicited two advantages: firstly, it allowed us to use lower reaction temperatures, which avoided thermal polymerization of the acrylates; secondly, it enabled us to use less alcohol, which was important with synthetically valuable monomers. The degree of substitution (DS) was controlled by the amount of THF/water azeotrope removed. DS values ranged from 25%-75% as evaluated by endcapping of the unreacted silanols with trimethylsilylchloride (TMSCl).[9]

Scheme IV

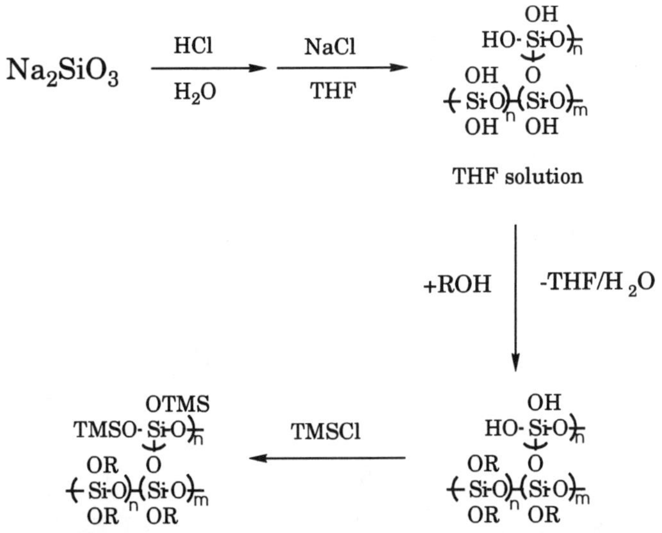

$$Na_2SiO_3 \xrightarrow[H_2O]{HCl} \xrightarrow[THF]{NaCl} \begin{array}{c} OH \\ HO\text{-}Si\text{-}O \end{array}_n$$

THF solution

+ROH | -THF/H$_2$O

$$\xleftarrow{TMSCl}$$

Composites were synthesized by allowing the silicic ester to condense in a solution of a polymerizable monomer and a free radical initiator (typically benzoyl peroxide) or aqueous ROMP catalyst (Scheme V). In most cases, transparent inorganic-organic composites are formed. A variety of cosolvent monomers can be used, including cross-linking agents, in order to achieve different properties in the final composite. For example, poly(hydroxyethylacrylate) (HEA) is a rubbery solid at room temperature (Tg=-15°C), whereas poly(hydroxyethylmethacrylate) (HEMA) is glassy at room temperature (Tg=55°C). A composite produced from a silicic ester of HEA is a rubbery material. Alternatively, using a HEMA silicic ester produces a stiff material. The glass/polymer ratio also influences the properties of the composite. Preliminary studies with (HEMA)/glass composites indicated that higher glass composition improves the hardness and compression modulus of the composites.

Conclusion

SIPN technology provides a convenient method for incorporating insoluble polymers into sol-gel composites. This approach has been extended towards the synthesis of non-shrinking sol-gel composites. The use of silicic acid esters allows

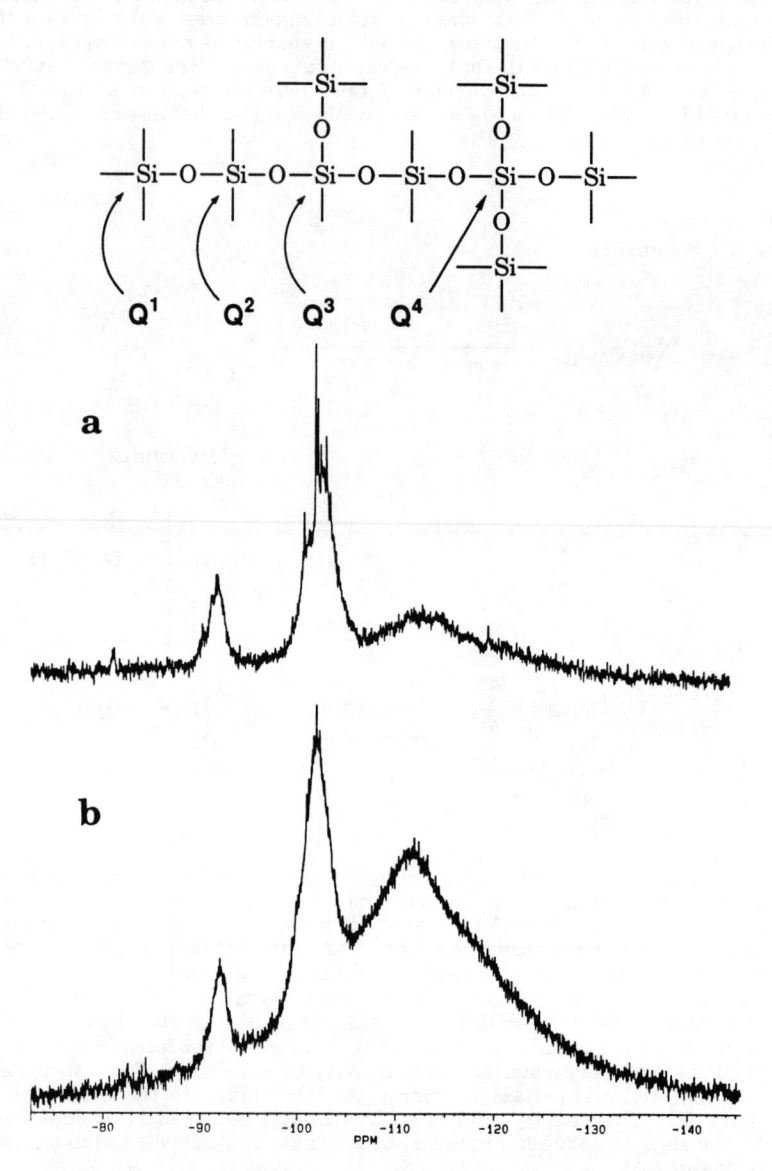

Figure 1. ^{29}Si NMR spectra of poly(silicic acid): a) 3M HCl, 1 hour reaction; b) 6M HCl, 1 hour reaction.

for the production of non-shrinking composites with a wide range of glass/polymer ratios. The physical and mechanical properties of these composites can be tailored by changing the glass content and using different monomers. Since long drying times are unnecessary, these composites may prove to be suitable for a wide range of applications.

Acknowledgements

We gratefully acknowledge financial support for this work from the National Science Foundation, Presidental Young Investigator Award; The Alfred P. Sloan Foundation; the Office of Naval Research; The Center for Advanced Materials Science Division, Lawrence Berkeley Laboratory; and Industrial Support. M.W.E. acknowledges the Department of Education for a Graduate Student Fellowship.

Scheme V

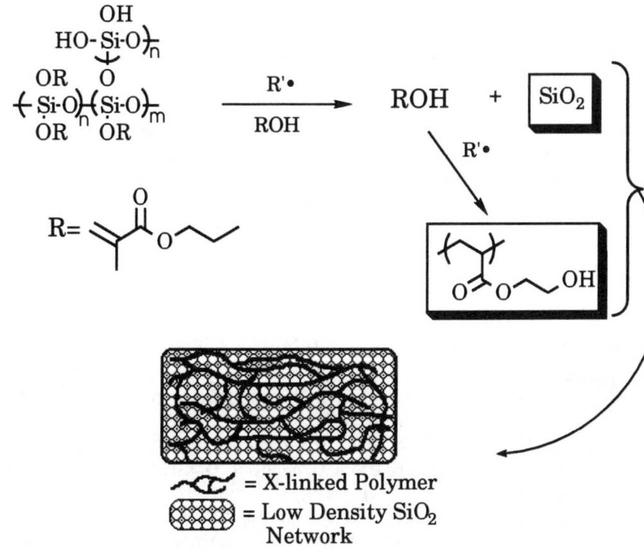

Experimental

SIPN composites using various monomers and TMOS. SIPN xerogels were prepared by weighing 5.0 mg of K_2RuCl_5 and 10-200 mg of monomer in separate scintillation vials. The solids were dissolved in 1.0 mL of aqueous NaF followed by 1.0 mL of methanol or ethanol. The NaF concentration is dependent upon monomer concentration and was chosen such that the gelation time did not exceed the time needed for complete polymerization of the ROMP monomers (see Table 1). Co-solvent is dependent upon the solubility of the ROMP polymer. 1.0 mL of TMOS was then added. The vial was sealed and placed in a 60 °C bath for the required amount of time. The gel was removed from the bath and allowed to set overnight. The gel was dried by stretching a thin sheet of parafilm over the mouth of the vial and allowing the solvent to diffuse slowly through the parafilm. After one week, the partially dried gels were submersed in water for 24 hours to remove residual Ru. The gels were then dried in the same manner for an additional three weeks to yield clear, low density glass composites.

Representative Non-Shrinking Composite Synthesis from Tetraalkeneyl orthosilicates: In a scintillation vial was combined 1.0 g of tetrakis(7-oxanorbornene methoxy) silane, 0.5 g of 7-oxanorbornene methanol, 5 mg of $K_2RuCl_5 \cdot H_2O$, and 0.01 mL of 50 mM NaF. The solution was heated at 60 °C under N_2 for 2 hours with occasional swirling during the first 5 minutes to maintain homogeneity. After one hour, a transparent, light orange, rubbery composite was obtained. Further heating at 100 °C for 24 hours produced a more rigid composite material. The same procedure was used for the free radical composite synthesis with benzoyl peroxide as the polymerization catalyst.

Representative Non-Shrinking Composite Synthesis from Silicic Esters: A 35 mL aliqout (1.7 g HEMA silicic ester) was taken from a solution containing 4.8 g of HEMA silicic ester in 100 mL of THF. To the aliqout solution was added 0.5 g of HEMA, 20 mg of ethylene diacrylate, and 20 mg of benzoyl peroxide. The solution was concentrated *in vacuo* until a viscous solution formed. The solution was poured into a scintillation vial and the vial was evacuated in order to outgas the sample before polymerization. After 2-3 hours, the vial was purged with N_2 and placed in a 60 °C bath for 3 hours. The solidified sample was removed by crushing the vial. The composite sample was then placed in a vacuum oven at 80 °C for 24 hours after which a 2.0 g transparent composite containing approximately 50% glass and 50% polymer was obtained. The same general procedure was used for the 7-oxanorbornene monomers with $K_2RuCl_5 \cdot H_2O$ as the polymerization catalyst. Analysis: Found C, 30.41%; H, 4.78%, SiO_2 residue, 48.9%.

References

1. a) R. K. Iler, The Chemistry of Silica, (Wiley, New York, 1955). b) Ultrastructure Processing of Advanced Ceramics, edited by J. D. Mackenzie and D. R. Ulrich (Wiley, New York, 1988). c) L. L. Hench and J. K. West, Chem. Rev. 90, 33 (1990).
2. a) B. Wang, G. L. Wilkes, C. D. Smith, and J. E. McGrath, Polymer Communications 23, 400 (1991). b) G. L. Wilkes, B. Orler, and H. Huang, Polym. Prep. 26, 300 (1985). c) H. Huang, B. Orler, and G. L. Wilkes, Polym. Bull. 14, 557 (1985). d) H. Huang, R. H. Glaser, and G. L. Wilkes, Polym. Prep. 28, 434 (1987). e) H. Huang, B. Orler, and G. L. Wilkes, Macromolecules, 20, 1322 (1987). f) H. Huang, G. L. Wilkes, and J. G. Carlson, Polymer, 30, 2001 (1989). g) R. H. Glaser, and G. L. Wilkes, Polym. Bull. 19, 51 (1988). h) A. B. Brennan, B. Wang, D. E. Rodriques, and G. L. Wilkes, J. Inorg. and Organomet. Polym. 1, 167 (1991).
3. a) H. Schmidt, J. Non-cryst. Solids, 112, 419 (1989). b) H. Schmidt and G. Phillip, J. Non-cryst. Solids, 63, 283 (1984). c) E. J. A. Pope and J. D. Mackenzie in Better Ceramics Through Chemistry II, edited by C. J. Brinker, D. E. Clark, and D. R. Ulrich (MRS, Pittsburg, PA, 1986). d) E. J. A. Pope, M. Asami, and J. D. Mackenzie J. Mater. Res. 4, 1018 (1989).
4. a) S. J. Clarson and J. E. Mark Polym. Commun. 28, 249 (1987). b) Y. P. Ning, M. Y. Tang, C. Y. Jiang, J. E. Mark, and W. C. Roth J. Applied Polym. Sci. 29, 3209 (1984). c) J. E. Mark, C. Jiang, M. Y. Tang, Macromolecules, 17, 2616 (1984). d) J. E. Mark, Y. P. Ning, M. Y. Tang, and W. C. Roth, Polymer, 26, 2069 (1985).

5. a) B. M. Novak, M. W. Ellsworth, T. I. Wallow and C. Davies Polym. Prep. 31, 698 (1990). M. W. Ellsworth and B. M. Novak, J. Am. Chem. Soc. 113, 2756 (1991). c) B. M. Novak and C. Davies, Macromolecules, 24, 5481 (1991).
6. J. Ebelman, Justus Liebigs Ann. Chem. 57, 331 (1846).
7. M. W. Ellsworth and B. M. Novak, Polym. Prep. 33, 1088 (1992).
8. Equation 2 is derived by noting that the percent glass in the final material is a function of the weight percent glass contributed by each silicon center (Q^2, Q^3, and Q^4) and that the amount of glass is reduced by the alkoxide substitution (two alkoxides per Q^2 center and one alkoxide per Q^3 center).
9. a) Y. Abe and T. Misono, J. Polym. Sci., Polym. Lett. Ed. 20, 205 (1982). b) ibid, J. Polym. Sci., Polym. Chem. Ed. 21, 41 (1983).
10. O. Girard, A. Guillermo, and J. P. Cohen Addad, Makromol. Chem. 30, 69, (1989).

REINFORCEMENT FROM IN-SITU PRECIPITATED SILICA IN POLYSILOXANE ELASTOMERS UNDER VARIOUS TYPES OF DEFORMATION

JAMES E. MARK, SHUHONG WANG, PING XU, AND JIANYE WEN, Department of Chemistry and the Polymer Research Center, The University of Cincinnati, Cincinnati, OH 45221-0172

ABSTRACT

Elastomeric networks prepared by tetrafunctionally end linking hydroxyl-terminated poly(dimethylsiloxane) chains (PDMS) were filled by the in-situ precipitation of silica. The resulting networks were investigated under uniaxial elongation, biaxial extension, shear, and torsion in order to characterize the resulting changes in mechanical properties. Compared with the unfilled networks, the silica-filled materials showed large reinforcing effects. Specifically, their values of the modulus, ultimate strength, and rupture energy increased significantly. The results thus indicate that the PDMS networks filled by the in-situ precipitation of silica have very good mechanical properties in several, rather different deformations. Examples of other deformations of interest are equilibrium swelling, and dynamic cycling for characterization of compression set.

INTRODUCTION

Poly(dimethylsiloxane) (PDMS) $[-Si(CH_3)_2O-]_x$ is a good example of an elastomer which requires considerable reinforcement from silica or some other filler before it is useful in most industrial applications.[1-3] Typically, such fillers are blended into a polymer of high molecular weight, prior to its being cross linked into the final network structure. This is a difficult, tedious procedure since the polymer usually has a very high bulk viscosity and the filler is typically badly agglomerated.[4] It is not surprising, therefore, that good dispersions frequently require a great deal of time, energy, and patience.

For these and other reasons, an alternative, novel approach to obtaining such reinforcement has been developed. It involves precipitating ceramic-type fillers such as silica (SiO_2) into a network by means of a sol-gel technique. The most important reaction of this type involves the catalytic hydrolysis of tetraethoxysilane (TEOS)[4-10]

$$Si(OC_2H_5)_4 + 2H_2O \longrightarrow SiO_2 + 4C_2H_5OH \qquad (1)$$

where the byproduct, ethanol, is easily removed because of its volatility. The particles generated in this manner have a narrow distribution of sizes, with most diameters typically in the 200 - 250 Å, and show little if any aggregation.[11,12] There have been numerous experiments carried out on such in-situ filled PDMS networks in uniaxial extension, and they have shown that these elastomers have good mechanical properties.[13-15] However, there is relatively little analogous data available for the other types of deformation such as biaxial extension (compression), shear, and torsion. This is true even for elastomers filled in the usual manner, by blending of particles prior to cross linking.[16] It would obviously be interesting and important to use a variety of deformations to characterize the effects of fillers on the mechanical properties of elastomers, particularly those reinforced by the new sol-gel technique.

The present article reports on some studies carried out to characterize the reinforcement of in-situ precipitated silica in PDMS elastomers in elongation, biaxial extension, pure shear, and torsion. These types of deformation are important in the characterization of elastomeric materials, but are not much studied because they are more difficult to impose than simple elongation (uniaxial extension). Comparisons among the results obtained should ultimately provide a general molecular understanding of the mechanical behavior and ultimate properties of filler-reinforced elastomers.[17]

PREPARATION OF NETWORKS

Hydroxyl-terminated PDMS samples were typically tetrafunctionally endlinked with a stoichiometrically equivalent amount of TEOS in Teflon® molds at room temperature, using 1 wt.% stannous-2-ethyl-hexanoate as catalyst.[15,18] The resulting networks were then extracted with a solvent, deswelled with a non-solvent, and finally dried. Typically, only small amounts of soluble (uncross-linked) polymer were thus removed.

IN-SITU PRECIPITATION OF SILICA

Pieces of PDMS sheets were typically immersed in TEOS for different periods of time in order to get different degrees of swelling. This was done in order to precipitate different amounts of silica into the networks.[17] After the swollen sheets were removed from the TEOS, they were placed into an aqueous solution of catalyst, and the hydrolysis reaction permitted to proceed at room temperature.[12] The shapes and sizes of the test samples depended, of course, on the type of deformation to be used to characterize the reinforcement.

MECHANICAL PROPERTY MEASUREMENTS

Elongation

The strips used for these measurements were cut from the same samples which were used for the biaxial extension measurements. These measurements were carried out in the usual manner,[15,18] by simply stretching the strips between two clamps, one of which was attached to a stress gauge, and measuring stretched and unstretched lengths with a cathetometer. These measurements, and the ones involving other deformations, were generally conducted in the vicinity of 25 °C.

The elastomeric quantities of primary interest obtained from these measurements are the stress f^* and the reduced stress $[f^*] \equiv f^*/[A_0(\alpha - \alpha^{-2})]$, where A_0 is the cross-sectional area of the undistorted sample, and $\alpha = L/L_i$ is its relative length.

Biaxial Extension

The equations for biaxial extension are identical to those for uniaxial elongation and, thus, the above-mentioned elastomeric quantities are used here as well.[17,19]

The method utilized to produce the desired equi-biaxial extension was inflation of a circular sheet clamped around its circumference, into the form of part of a spherical balloon. The state of strain corresponds exactly to that produced by uniaxial compression, the only difference being in the nature of the applied stress.

A detailed description of this type of deformation is described elsewhere.[17,19] The system for measuring the biaxial extension is primarily composed of a cell, micrometer slide cathetometer, travelling microscope, mercury manometer and fluidized temperature-controlled bath. The measurement techniques are described elsewhere.[17,19] The network sheet for biaxial extension was placed between the base of the cell and a cover, which was held in place by four screws. Water which was free of dissolved air was forced into the cell by increments. The pressure required was obtained from the heights of mercury in the manometer arms. The strain was obtained from the vertical distances between the points marked on the inflated sheet (as measured with a cathetometer), and from the horizontal distances (measured with a travelling microscope).

Shear

Pure shear involves extension in three perpendicular directions without rotation of the principal axes of the strain. In this work, it was achieved by stretching a rectangular sheet in one direction.[16] The extension ratio in this direction is α, and the perpendicular or transverse direction remains unchanged, to keep $\alpha_2 = 1$. This is achieved by having the width of the test sheet very much greater than its length, which makes changes of the sheet width negligible. The stress t_2 is automatically generated as a result of the restraints introduced by the clamps. The principal stress t_1 and its extension ratio α can then be measured. The equilibrium tensile force can be read in digital form from an amplifier. The apparatus for these measurements is also described in detail elsewhere.[17]

Torsion

The apparatus for the torsion measurements was based on Gent's modification[20] of the equipment described by Treloar.[16] For this type of deformation, the shear modulus is calculated from $G = 2M/\pi\phi a_0^4$, where M is the torsion couple, ϕ the torsion in radians per unit length of the strained axial length, and a_0 the unstrained radius. Sample lengths were approximately 35 mm, and rotations ranged from approximately 130 to 1300 °.[21]

Equilibrium Swelling

In a swelling experiment, the network is typically placed into an excess of solvent, which it imbibes until the dilational stretching of the chains prevents further absorption of solvent.[15,16,22] Adsorption of some network chains onto the particles of a reinforced elastomer would, of course, reduce the extent of swelling at equilibrium. Characterization of the extent of such equilibrium swelling as a function of filler constitution, filler amount, and the method of generating and incorporating it into the elastomer are the issues of greatest interest with regard to this type of deformation.

Cyclic Deformations (Compression Set)

It is also of considerable interest to see how filled elastomeric materials perform in cyclic deformations, in order to characterize their resistance to creep or to failure by fatigue. In a typical experiment of this type, silica-filled PDMS elastomers are being subjected to cyclic compressions, with the modulus and sample dimensions monitored as a function of the number of cycles, the time period of cycling, or the power consumed.[23]

RESULTS AND DISCUSSION

Elongation and Biaxial Extension

Since biaxial extension is equivalent to uniaxial compression, both types of data can be combined in the same plot. In this way, a full spectrum of stress-strain data for both elongation and compression can be viewed, and the behavior of networks in the two deformation regions can be compared directly. Results obtained for both elongation ($\alpha^{-1} < 1$) and compression ($\alpha^{-1} > 1$) are therefore depicted conjointly in Figure 1, where the modulus is plotted against α^{-1}. It is obvious that very strong reinforcing effects from the precipitated silica have been obtained. This is evident from both the large upward shifts of the isotherms as a whole and from the pronounced upturns at both high elongations ($\alpha^{-1} < 1$), and at high compressions (biaxial extensions) ($\alpha^{-1} > 1$). As was already mentioned, such desired reinforcing effects do not occur for the reference (unfilled) sample.

It is interesting to note that the extent of the reinforcemnent, as gauged by the magnitudes of the upturns, is approximately the same in uniaxial and biaxial extension. The range of deformation over which it occurs, however, seems to be larger in the case of the biaxial extension. The pronounced maxima and minima in the isotherms for the filled elastomers in biaxial extension is not understood at present, and represent challenges to theories on the origin of reinforcing effects in elastomers in general.[17]

The effect of filler content on the rupture moduli in elongation and in compression is shown in Figure 2. The dependences are seen to be quite similar, with the curves for uniaxial and biaxial extension being very nearly parallel to one another. Figure 3 shows the effect of filler content on the deformation at rupture in elongation and in compression. The two behaviors are again quite similar, in that increase in filler content changes these values of α_r in the direction of smaller deformations ($\alpha \rightarrow 1$).

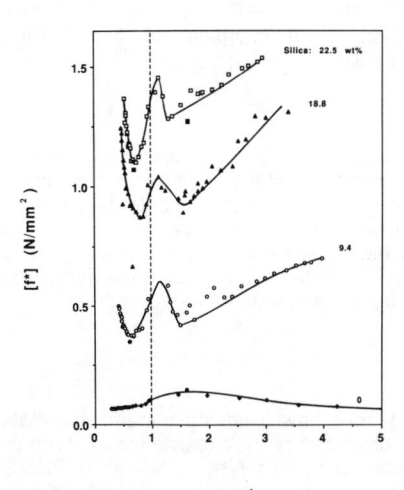

Fig. 1. Stress-strain isotherms in elongation (region to the left of the vertical dashed line, with $\alpha^{-1} < 1$), and in compression (to the right, $\alpha^{-1} > 1$).[17] The filled points represent the data used to test for reversibility.

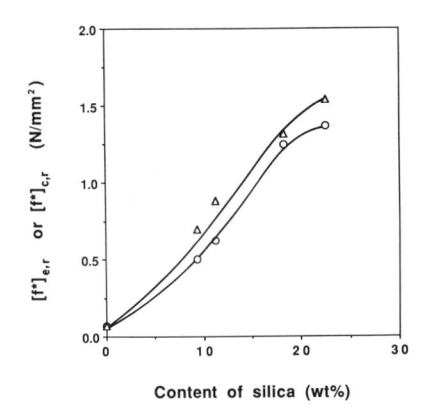

Fig. 2. Effect of filler content on the rupture modulus in elongation $[f^*]_{e,r}$ (O), and in compression $[f^*]_{c,r}$ (△).[17]

Similarly, values of the energy E_r required for rupture in both uniaxial and biaxial (compression) extension can be obtained by direct integration of $\int_\alpha f^* d\alpha$. For both deformations, the increases in $[f^*]_r$ predominate over the decreases in α_r, and E_r increases substantially with increase in filler content.[17] The two curves lie nearly parallel to one another, in a way similar to that shown in Figure 2.

Shear

The stress-strain isotherms obtained for both the unfilled and the in-situ filled PDMS networks, represented in terms of the pure shear modulus G and principal extension ratio α, are shown in Figure 4. The results in simple shear, in terms of

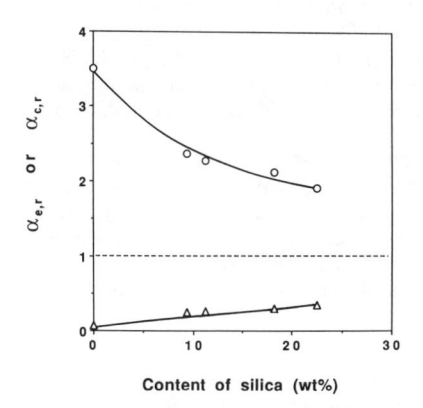

Fig. 3. Effect of filler content on the deformation at rupture in elongation $\alpha_{e,r}$ and in compression $\alpha_{c,r}$.[17] See legend to figure 2.

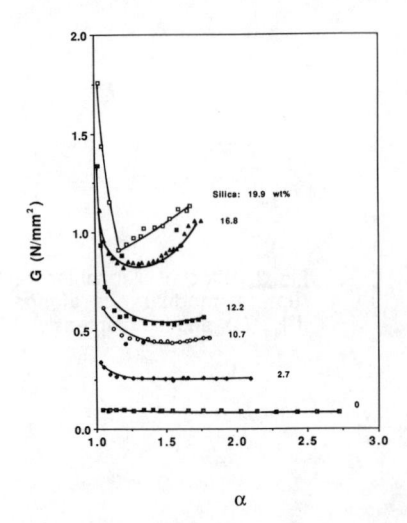

Fig. 4. Stress-strain isotherms in pure shear in terms of the modulus. The filled points represent the data used to test for reversibility.[17]

the shear strain γ, were essentially identical and are presented elsewhere.[24] As in the case of uniaxial extension and compression, the addition of filler shifts the isotherms upwards to significantly higher vaues of the modulus. In the case of the filled networks there is an initial decrease in the modulus with increase in deformation. Also, the larger the amount of filler present, the more pronounced the decrease. This may be due to stress-induced rearrangements of the chains in the vicinity of the filler particles.[17] Also of possible relevance is the fact that increase in amount of filler decreases the number of load-bearing chains passing through unit cross-sectional area, and changes the distribution of their end-to-end distances.

From these results, it can readily be seen that the filled samples have much higher shear moduli than the corresponding unfilled sample. Also, at high deformation, there are pronounced upturns in the reduced stress or modulus in the case of the filled samples, which indicates very good reinforcement. The magnitudes of the upturns are much less pronounced, however, than they were found to be in uniaxial extension and compression. In any case the extension ratio α or shear strain γ at which the upturn appears is seen to decrease with increase in wt % filler. As is usually the case,[1-3] the maximum deformability decreases as well. Clearly, the behavior observed here is very similar to that described above for uniaxial and biaxial extension.

Also of interest is the dependence of the ultimate properties on the silica content.[17] It was observed that α_r and γ_r decrease with increase in silica content, while G_r and f^*_r increase at lower contents, but appear to be leveling off or even decreasing at higher contents. This is presumably because of the decreasing extensibility. The rupture energy E_r depends on silica content in a way very similar to that of G_r and f^*_r. The dependences are very similar to those described in the preceeding Section.

Torsion

Some prelimary isotherms obtained on PDMS-SiO$_2$ in-situ filled elastomers in torsion[21] are shown in Figure 5. As was the case in the previous deformations, increase in filler content shifts the isotherms upward, to significantly

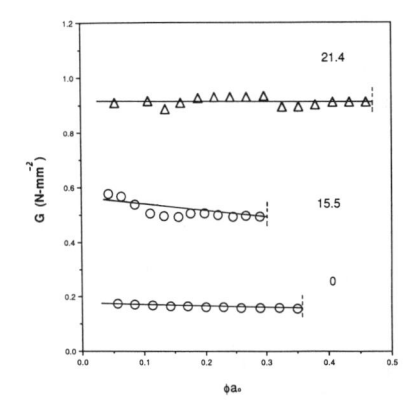

Fig. 5. Shear modulus shown as a function of the torsion strain expressed as ϕa_0 (radians), where ϕ is the twist in radians per unit length of the strained axis and a_0 is the radius of the unstrained (cylindrical) sample.[21] The vertical dashed lines locate the rupture points.

higher values of the shear modulus. These curves are significantly different from the previous ones, however, in that there are no upturns in the modulus at high deformations. This could be due to either the nature of the deformation or to the inability, to date, of reaching sufficiently high deformations.

Equilibrium Swelling

Although this is a potentially interesting area, results to date are insufficient to lead to any useful conclusions for this type of deformation.

Cyclic Deformations (Compression Set)

Experiments are now being carried out on a variety of PDMS-SiO$_2$ elastomers,[23] and results, particularly with regard to compression set, should be forthcoming in the near future.

CONCLUDING COMMENTS

The results presented here are interesting in that they demonstrate that elastomeric networks filled by the in-situ precipitation of silica have good mechanical properties, not only in uniaxial extension, but also in compression, shear and torsion. Other types of deformation, for example swelling equilibrium and cyclic deformations, are in progress.

ACKNOWLEDGEMENT

It is pleasure to acknowledge the financial support provided by the Army Research Office through Grant DAAL03-90-G-0131, and by the National Science Foundation through Grant DMR 89-18002 (Polymers Program, Division of Materials Research). We are also very grateful for the P. J. Flory Memorial Fellowship awarded P. Xu by the Rubber Division of the American Chemical Society, and an equipment grant obtained from the GenCorp Foundation through Dr. Russell A. Livigni.

REFERENCES

(1) Boonstra, B. B. *Polymer* **1979**, *20*, 691.
(2) Warrick, E. L.; Pierce, O. R.; Polmanteer, K. E.; Saam, J. C. *Rubber Chem. Technol.* **1979**, *52*, 437.
(3) Enikolopyan, N. S.; Fridman, M. L.; Stalnova, I. O.; Popov, V. L. *Adv. Polym. Sci.* **1990**, *96*, 1.
(4) Mark, J. E.; Pan, S.-J. *Makromol. Chemie, Rapid Comm.* **1982**, *3*, 681.
(5) Mark, J. E. *CHEMTECH* **1989**, *19*, 230.
(6) Wilkes, G. L.; Brennan, A. B.; Huang, H.-H.; Rodrigues, D.; Wang, B. In *Polymer-Based Molecular Composites*; Schaefer, D. W.; Mark, J. E., Ed.; Materials Research Society: Pittsburgh, 1990; Vol. 171; p 15.
(7) Schmidt, H. In *Polymer-Based Molecular Composites*; Schaefer, D. W.; Mark, J. E., Ed.; Materials Research Society: Pittsburgh, 1990; Vol. 171; p 3.
(8) *Polymer-Based Molecular Composites*; Schaefer, D. W.; Mark, J. E., Ed.; Materials Research Society: Pittsburgh, 1990; Vol. 171.
(9) Mark, J. E.; Schaefer, D. W. In *Polymer-Based Molecular Composites*; Schaefer, D. W. and Mark, J. E. Ed.; Materials Research Society: Pittsburgh, 1990; Vol. 171; p 51.
(10) Mark, J. E. *J. Appl. Polym. Sci., Appl. Polym. Symp.* **1992**, *50*, 273.
(11) Mark, J. E.; Ning, Y.-P.; Jiang, C.-Y.; Tang, M.-Y.; Roth, W. C. *Polymer* **1985**, *26*, 2069.
(12) Xu, P.; Wang, S.; Mark, J. E. In *Better Ceramics Through Chemistry IV*; Zelinski, B. J. J., Brinker, C. J., Clark, D. E., and Ulrich, D. R., Ed.; Materials Research Society: Pittsburgh, 1990.
(13) Mark, J. E. *Brit. Polym. J.* **1985**, *17*, 144.
(14) Ning, Y.-P.; Mark, J. E. *Polym. Eng. Sci.* **1986**, *26*, 167.
(15) Mark, J. E.; Erman, B. *Rubberlike Elasticity. A Molecular Primer*; Wiley-Interscience: New York, 1988.
(16) Boyce, P. H. and Treloar, L. R. G. *Polymer* **1970**, *11*, 21.
(17) Wang, S.; Xu, P.; Mark, J. E. *Rubber Chem. Technol.* **1991**, *64*, 746.
(18) Mark, J. E.; Sullivan, J. L. *J. Chem. Phys.* **1977**, *66*, 1006.
(19) Xu, P.; Mark, J. E. *Rubber Chem. Technol.* **1990**, *63*, 276.
(20) Gent, A. N.; Kuan, T. H. *J. Polym. Sci., Polym. Phys. Ed.* **1973**, *2*, 1723.
(21) Wen, J.; Mark, J. E., unpublished results.
(22) Erman, B.; Mark, J. E. *Ann. Rev. Phys. Chem.* **1989**, *40*, 351.
(23) Fitzgerald, J. J.; Wen, J.; Mark, J. E., unpublished results.
(24) Wang, S. *Ph. D. in Chemistry, University of Cincinnati* **1991**.

MOLECULAR WEIGHT DEPENDENCE OF DOMAIN STRUCTURE IN SILICA-SILOXANE MOLECULAR COMPOSITES

Tamara A. Ulibarri, Greg Beaucage, Dale W. Schaefer, Bernard J. Olivier and Roger A. Assink

Sandia National Laboratories, Albuquerque, NM 87185

ABSTRACT

A detailed investigation of the molecular weight dependence of silica growth in *in situ* filled polydimethylsiloxane/tetraethylorthosilicate (PDMS/TEOS) materials was conducted using small angle neutron scattering (SANS). Composite materials were produced by using TEOS to simultaneously produce the glassy filler phase and to crosslink linear, hydroxyl terminated PDMS of variable molecular weight, M. Correlated domains of glassy filler were produced. The morphology of the *in situ* filled material showed a definite dependence on the molecular weight of the precursor polymer. The scale, R, of the glassy domains follows de Gennes' description of phase separation in a crosslinked system (R α $M^{1/2}$).

INTRODUCTION

The use of sol-gel techniques to produce inorganic/organic composite materials dispersed on the nanometer level has recently become an area of great activity [1-8]. We have previously reported scattering studies of *in situ* filled PDMS/TEOS systems [3]. These studies demonstrated that a variety of filler structures can be obtained, but failed to establish the essential factors leading to the observed morphologies. Like their solution polymerized analogs [9] the structure of *in situ* polymerized silicates can reflect both near equilibrium processes (e. g., spinodal decomposition) and kinetic processes (e. g., reaction limited aggregation). In view of the complexity of these systems, a thorough understanding of the relationships between the synthetic protocol, structure and properties is required to exploit the potential of these novel materials.

Prior scattering data were analyzed using a fractal approach in which the high "q" (q=4π/λ sin(θ/2)) Porod regime is used to infer the structure [3]. This approach is analogous to the one we used to analyze conventional fumed silica, the current filler of choice for siloxane elastomers [9]. Depending on the length scale, fumed silica displays structures ranging from mass-fractal type branched polymers (Porod slope > -3) through diffuse interfaces (Porod slope < -4) and compact colloidal particles (Porod slope = -4). The generation of these structures depends on processing parameters.

Previous work involving the PDMS/TEOS system [10] only briefly explored the effect of oligomer molecular weight and concluded that phase separation was less pronounced with lower molecular weight polymers (a result we confirm). Additional studies exploring molecular weight effects in poly(tetramethylene oxide) [PTMO]/[TEOS] systems [8a, 11], found an increase in the correlation distance with increasing PTMO molecular weight (i.e., an increase in the interdomain spacing for one component in a two-phase system).

In this paper we further investigate the molecular weight relationship by carrying out a detailed study on a PDMS/TEOS system and analyzing our SANS data with respect to de Gennes' description of phase separation in a crosslinked system [12]. Completing this evaluation of the growth process should allow us to predict more accurately the morphology as a function of processing parameters.

EXPERIMENTAL

Materials

Hydroxyl-terminated PDMS (Petrarch/Huls America) having a number-average molecular weight of 400-700; 2,000; 4,200; 18,000; 36,000; 77,000; 150,000; 310,000 g mol^{-1}, tetraethylorthosilicate (Aldrich) and stannous 2-ethyl hexanoate (Pfaltz & Bauer) were placed in a glovebox and used as received. In the glovebox, each respective PDMS was mixed thoroughly with 43.3 wt % TEOS to yield a PDMS/TEOS mixture weighing approximately 7g. After mixing, stannous 2-ethyl hexanoate was added at catalyst/TEOS ratios of 1/300 and 1/70. All reactions were run in an undiluted state except PDMS = 150,000 and 310,000 g mol^{-1}. Due to the high viscosity of these polymers, these reactions were diluted with 4 mL of either toluene (1/300 cat. ratio) or tetrahydrofuran (1/70 cat. ratio). All of the mixtures appeared to be homogeneous. After mixing was complete, the reactions were poured into molds yielding thicknesses of 1.0 - 1.5 mm. The filled molds were then placed in a 31% R_H chamber and the reactions were allowed to cure at room temperature for 1 week. The resulting materials were then leached with a 1:1 tetrahydrofuran/isopropanol mixture and allowed to air dry.

Instrumentation and Methods

Neutron scattering was performed at the Manuel Lujan Neutron Scattering Center at Los Alamos National Laboratory, at the Cold Neutron Research Facility at the National Institute of Standards and Technology and at the High Flux Isotope Reactor at Oak Ridge National Laboratory. Data from different q ranges were matched by an arbitrary vertical shift factor.

The ^{29}Si NMR spectra were recorded at 39.6 MHz on a Chemagnetics console interfaced to a General Electric 1280 data station and pulse programmer. The samples were spun about the magic angle at 5 to 6 kHz. A direct polarization sequence was used with pulse delay times of 240s (a factor of 5 times the longest observed T_1) so the resonance areas are expected to be quantitative.

Sample density values were determined using a density gradient column [13]. Modulus profiles were obtained using a thermomechanical analyzer which has been modified to give tensile compliance measurements which are inversely related to the more commonly measured tensile modulus [14]. For most of the samples, the profiles indicate that the deposition of silica is not homogeneous and is more concentrated on the surface of the sample than the "bottom."

MODEL

The *in situ* filling procedure used here leads to simultaneous silica formation and polymer crosslinking. Therefore, the phase separation of the hydrolyzed TEOS, which is driven by the extent of the hydrolysis and condensation, occurs in a multicomponent, partially crosslinked polymeric network leading to a complicated interplay of chemical growth and phase separation.

The simplest model for phase separation in the presence of a network has been developed by de Gennes [12]. In de Gennes' approach the terms in the free energy which drive the phase separation are opposed by a Cahn-Hillard concentration gradient term and by a term which reflects the modulus of the network derived from rubber elasticity theory. The size scaling of the domain structure is predicted by using the continuity equation and Fourier transformation of the free energy expression. Due to the rubber elasticity term, this size scale is proportional to the distance between crosslinks for the elastomer network, which scales as the square root of the degree of polymerization.

de Gennes' theory neither postulates nor predicts any particular structure for the phase separated system, but rather describes only the behavior of the characteristic length of the resulting domains. In the case of thermal phase separation one might expect a spinodal type structure if the system is driven deeply into the incompatible regime. In a chemically driven phase separation between an inorganic glass and a network polymer, however, a myriad of structures is possible (as is evidenced in data previously presented [3]).

In this study, we have used a general data analysis procedure that includes a degree of flexibility in terms of the interfacial structure, the interdomain correlations and the domain size. This expression was found to be the simplest physically reasonable model capable of fitting the data (Fig. 1). Models based on a single length scale (domain size = correlation length), for example, cannot fit the data. Although the model is based on "particles," the conclusions derived from it should be valid when applied to spinodal patterns as well. This generality derives from the fact that we are primarily interested in the length scale of the domains.

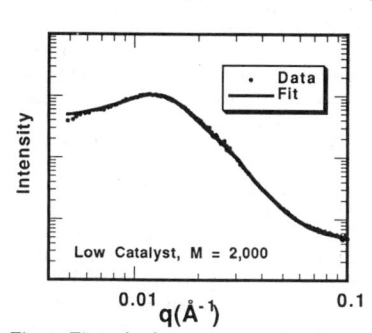

Fig. 1. Fit to the data using equations 1-4.

In the model, the scattered intensity is described by a contrast factor A, a background term B, a function describing the correlations of the filler particles S(q) and a form factor for the particles F(q),

$$I(q) = A\, F(q)^2\, S(q) + B \qquad (1)$$

Filler particles are correlated according to a damped spherical model in which p is related to the number of coordinated filler particles (large p = higher number of particles within an effective coordination distance; p generally varies from 0 - 6) and 2ξ is the average correlation distance for the filler particles,

$$S(q) = \frac{1}{(1+p\,\Theta)} \qquad (2)$$

$$\Theta = 3\frac{(\sin(2q\xi) - 2\cos(2q\xi))}{(2q\xi)^3} \qquad (3)$$

The particle form factor F(q) is simply assumed to show large q scaling as a power law with the exponent m and the particulate "radius" R,

$$F(q)^2 = \frac{1}{(1 + 0.22\,(qR)^m)} \qquad (4)$$

The factor of 0.22 (not of interest here), was chosen to match smoothly the form factor of a sphere when qR = 1 and m = 4 [15] leading to an approximate identification of R with the particle radius of gyration. As was discussed above, power law slopes less than -4 indicate the presence of diffuse interfaces, which one might expect if partially polymerized TEOS dominated the particle surfaces. Fig. 2 shows a typical structure described by the above equations.

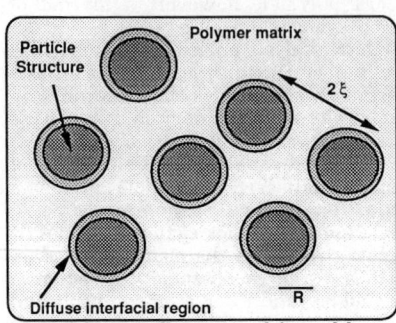

Fig. 2. Schematic illustration of the model embodied in equations (1-4).

RESULTS/DISCUSSION

Neutron scattering data for a series of elastomer molecular weights are shown in Fig. 3. The shift of the scattering profiles to a smaller q with M is consistent with the idea that elasticity controls the domain size.

The data from Fig. 3 generated a Porod regime with slopes between -4 and -5, consistent with a diffuse interface. A diffuse interface would be expected if partially reacted TEOS was present in the system. To verify the presence of partially hydrolyzed TEOS, the samples were inspected by ^{29}Si NMR. In this way, calculated densities and hydrolysis product ratios were derived. The ^{29}Si NMR data indicates that the TEOS has not been completely converted to pure SiO_2. In addition, experimental density measurements of the materials (from density gradient columns) agreed with the density values obtained by NMR, further indicating the presence of partially reacted TEOS.

In all cases, the correlation number, p, was close to 1 indicating weak correlations. The similarity of the packing factor and the surface characteristics indicates that the *in situ* reaction procedure used in this study produces similar structures differing only in characteristic length. This conclusion is independent of the validity of the correlated particle model used to extract the parameters.

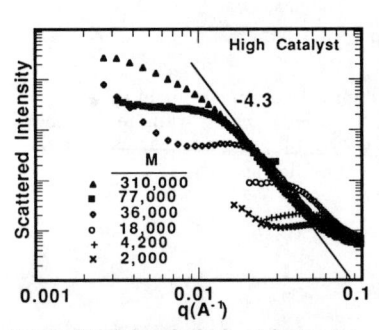

Fig. 3. SANS data for high catalyst series.

The domain size R was extracted from least-squares fits (e. g., Fig. 1) and is plotted in Fig. 4 versus elastomer molecular weight for both of the catalytic conditions used in this study. The observed slopes of 0.43 and 0.45 ± 0.10 are consistent with de Gennes' prediction of 0.5. Therefore, this study provides an indication of the essential physics underlying the structure of *in situ* composites.

High and low catalyst systems show identical behavior apart from a prefactor shift to larger length scales for the low-catalyst materials. Such a shift indicates a lower effective modulus of the rubbery phase at low catalyst amounts which is consistent with the occurrence of chain extension of the oligomers during crosslinking. Evidence for enhanced chain extension at low catalyst amounts has been previously noted [16].

Fig. 5 shows that the correlation range, ξ, also exhibits monotonic dependence on M. Although this parameter can be reliably extracted only at intermediate molecular weights, Fig. 5 still demonstrates that ξ approaches the "particle" size R as the domains become large. It is reasonable that the two lengths would show identical scaling at large R where packing constraints would force ξ to approach R.

It is also reasonable that the silica-like crosslinks would serve as nucleation sites for domain growth. If one assumes that each crosslink draws TEOS monomer from a volume that scales with the distance between crosslinks, one finds that the domain size should scale as $M^{1/2}$, identical to the prediction based on elasticity. We are now conducting experiments to distinguish between the two models, although the kinetic data suggests that elasticity is the dominant factor [3b].

Fig. 4. Dependence of the domain size on molecular weight of the elastomer precursor for high and low catalyst systems (slope given).

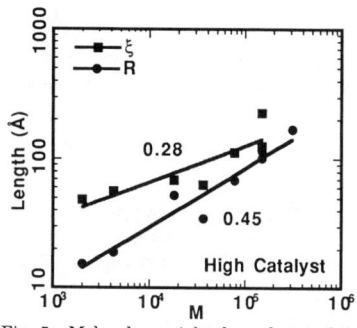

Fig. 5. Molecular weight dependence of the correlation range (ξ) and domain size (R) at high catalyst (slope given).

CONCLUSION

Controlled reaction conditions were used to produce *in situ* filled siloxane elastomers in a one-step process. The structure of the resulting two-phase system was explored using scattering techniques. The measured domain size follows de Gennes' prediction for phase separation in a crosslinked system. In all cases, the structure of the *in situ* filler shows diffuse interfaces that may be related to the presence of partially reacted TEOS.

ACKNOWLEDGEMENTS

We thank Dixie Harvey for technical assistance with scattering work at Oak Ridge National Laboratory. Small angle neutron scattering experiments were conducted with the supervision and assistance of Phil Seeger, George Wignall and Charlie Glinka. We thank M. Daoud for useful discussions. This work is sponsored by the U.S. Department of Energy under contract number DE-AC04-76DP00789.

REFERENCES

1. Due to space limitations, references 2-8 are only representative of the extensive work that has been done in this area.

2. Jim Mark, Chemtech, 19(4), 230 (89).

3. (a) D. W. Schaefer, J. E. Mark, D. W. McCarthy, L. Jian, C.-Y. Ning and S. Spooner, in "Ultrastructure Processing of Ceramics, Glasses and Composites", Ed. D. R. Ulrich and D. R. Uhlman, John Wiley & Sons, New York, 1992. (b) D.W. Schaefer, J. E. Mark, D. McCarthy, L. Jian, C.-C. Sun, B. Farago, Mat. Res. Soc. Symp. Proc., 171, 57 (1990). (c) D. W. Schaefer, L. Jian, C. -C. Sun, D. McCarthy, C. -J Jiang, Y. -P Ning and J. E. Mark, Polym. Prepr. (Am. Chem. Soc., Div. Polym. Sci.) 30(2) 102 (1989).

4. M. Spinu, C. Arnold and J. E. McGrath, Polym. Prepr. (Am. Chem. Soc., Div. Polym. Sci.) 30(2) 125 (1989).

5. M. W. Ellsworth and B. M. Novak, Polym. Prepr. (Am. Chem. Soc., Div. Polym. Sci.) 33(1) 1088 (1992).

6. (a) H. Schmidt, Mat. Res. Soc. Symp. Proc., 171, 3 (1990). (b) H. Schmidt and H. Wolter, J. Non-Cryst. Solids, 121, 428(1990).

7. Y. Wei, R. Bakthavatchalam, D. Yang and C. K. Whitecar, Polym. Prepr. (Am. Chem. Soc., Div. Polym. Sci.) 32(3) 503 (1991).

8. (a) G. L. Wilkes, H. Huang and R. H. Glaser in Silicon-Based Polymer Science (Advances in Chemistry Series 224); J. M. Zeigler and F. W. Feardon, Eds.; American Chemical Society: Washington, DC, 1990; pp 207-226. (b) B. Wang, G. L. Wilkes, J. C. Hedrick, S. C. Liptak and J. McGrath, Macromolecules, 24, 3449 (1991).

9. (a) D. W. Schaefer, Science, 243, 1023 (1989). (b) D. W. Schaefer and A. J. Hurd, Aerosol Sci. Technol., 12, 876 (1990).

10. H. Huang, B. Orler and G. L. Wilkes, Macromolecules, 20, 1322 (1987).

11. (a) D. E. Rodrigues, A. B. Brennan, C. Betrabet, B. Wang and G. L. Wilkes, Polym. Prepr. (Am. Chem. Soc., Div. Polym. Sci.) 32(3), 525 (1991). (b) H. Huang, G. L. Wilkes and J. G. Carlson, Polymer, 30(11), 2001 (1989). (c) H. Huang, R. H. Glaser and G. L. Wilkes, ACS Symposium Series 360, 354 (1988). (d) H. Huang and G. L. Wilkes, Polym. Prepr. (Am. Chem. Soc., Div. Polym. Sci.) 28(2), 244 (1987).

12. (a) P.G. de Gennes, J. de Phys., 40, 69 (1979). (b) A. Bettachy, A. Derouiche, M. Benhamou, and M. Daoud, J. Phys. I, 1, 153 (1991).

13. K. T. Gillen, R. L. Clough and N. J. Dhooge, Polymer, 27, 225 (1986).

14. K. T. Gillen, R. L. Clough and C. A. Quintana, Polym. Deg. Stab., 17, 31 (1987).

15. D. Posselt, J. S. Pederson and K. Mortensen, J. Non-Cryst. Solids, in press.

16. H. Huang, R. H. Glaser and G. L. Wilkes, Polym. Prepr. (Am. Chem. Soc., Div. Polym. Sci.) 28(1), 434 (1987).

REINFORCEMENT IN SILICONE ELASTOMERS. A SHORT REVIEW

JOHN C. SAAM*
*Michigan Molecular Institute, 1910 West St. Andrews Road, Midland MI, 48640

ABSTRACT

Selected topics on the use of particulate solids to build modulus and strength in silicone elastomers are briefly reviewed. Included are recent views advanced to explain the reinforcement phenomena, some more current approaches to particulate reinforcement and recommended areas for future endeavor.

INTRODUCTION

Properties of polydimethylsiloxane (PDMS) that give special characteristics to silicone elastomers originate from its flexible, open chain conformation, its weak intermolecular forces and a relative inertness to extreme environmental conditions. As a consequence the silicone elastomer system is unique in exhibiting extremely low glass and melting transitions, low bulk viscosity and viscosity-temperature coefficient, a low solubility parameter and surface energy and a relatively high permeability to gases [1]. In strained elastomeric networks, however, stress-crystallization is absent at ambient temperatures and restoring forces are therefore largely entropic. As a consequence pure PDMS networks are inherently weak and show little resistance to straining, tearing or abrading since mechanisms which can dissipate applied strain energy are absent [2]. Mechanisms must then be introduced in practical applications where mechanical stress is applied.

Strategies commonly applied to enhance strength in an elastomer all center about mechanisms which dissipate the energy before an advancing crack tip in the strained cross-linked elastomer matrix [3]. These include reducing the temperature of the test to increase the internal friction, the use of special network design to provide a bimodal distribution of cross-links [4, 5], coupling uniformly dispersed crystalline or thermoplastic domains to the network and, prior to cross-linking, adsorption of the precursor polymer onto dispersed particulate solids of very small size or high surface area. All have been applied to silicone elastomers, but by far the most important has been reinforcement with particulate solids, especially with the pyrogenic and precipitated silicas. Extant reviews on this type of reinforcement, while comprehensive, do not take into account the more current developments [2,6]. The present review selects some of the more recent advances while including some older references germane to the discussion.

GENERAL CONSIDERATIONS

Characteristics of particles especially suited for reinforcement in elastomers are a high specific surface area, controlled interaction of the surface with the polymer, bound polymer, structure, or a degree of aggregation of primary particles which persists through processing, and direct or indirect particle-particle interactions. These characteristics, important in silicone rubber as well as elastomers in general, are the subject of numerous publications and are reviewed in references 2 and 7. Despite the number of views advanced attempting to explain the reinforcement phenomena [8], a comprehensive, understandable theory explaining reinforcement is still lacking which can take into account all of the key observations and make quantitative predictions. This is not surprising in view of the complexity and wide range of the interactions in reinforced elastomers which lead to difficulties in devising and producing meaningful experimental observations.

The Extended Bound Polymer Network.

The importance of bound polymer in silica reinforced silicone elastomers became apparent beginning with the work of Southwart who found that unvulcanized PDMS reinforced with a pyrogenic silica contained large amounts of insoluble bound polymer which behaved like an elastomer by showing a high degree of elastic recovery. He sought to correlate observed increases in its Young's modulus, E, with the volume fraction of filler, ϕ, through the Guth-Gold equation:

$$E = E_0(1 + 2.5\phi + 14.1\phi^2) \tag{1}$$

Where E_0 is the Young's modulus when $\phi = 0$. Correlations were obtained only if ϕ included the volume of total bound rubber as well as the silica [9]. This would indicate that at low strains the effect of reinforcement on modulus could be explained in terms of the hydrodynamics of the reinforcing particle encapsulated by the adsorbed polymer. However, the upper limit of ϕ was only about 0.3 where equation 1 was valid. Also, it will be shown below that the same data can be fit to at least one other power law correlation having an entirely different interpretation.

In other work it was shown that a large part of the retractive force in PDMS elastomers was due to increases in the internal energy arising from the silica reinforcement. This was conjectured to be due to the large amount of energy required to deform the bound polymer-silica agglomerates [10]. More recently such structures were invoked to explain the dynamic rheological response of PDMS elastomers reinforced with several different pyrogenic silicas [11]. As proposed earlier to explain the effects of carbon black in natural rubber [12], agglomerates of PDMS and silica were thought to form extended networks [11].

Transmission electron microscopy confirmed that PDMS containing dispersed pyrogenic silica over a critical amount appeared to form a continuous network [13]. The network consisted of agglomerated silica apparently bound by adsorbed polymer that bridged filler particles. The critical amount of filler required to form a network corresponded approximately with the inflection points noted in plots of dynamic storage modulus or electrical conductivity vs. filler loading. The same critical point also corresponded with maxima in plots of filler loading vs elongation at break in the cured elastomer. This suggested behavior characteristic of a percolating network where the dynamic storage modulus, G', could be described by:

$$G' \sim (\phi - \phi_c)^t \tag{2}$$

where t is a critical exponent (0.85 for Aerosil 150) and ϕ_c is the observed threshold volume fraction of filler (0.091 for Aerosil 150) where G' shows an inflection point and sharply increases. Consistent with an expected reduction in polymer-filler interactions, treatment of the silica surface with trimethylsilyl groups to remove the free silanols gave a less defined ϕ_c shifted to higher values. This reviewer found that equation 2, with $\phi_c = 0.08$ and $t = 0.65$, fits Southwart's data [9] as well if not better than equation 1.

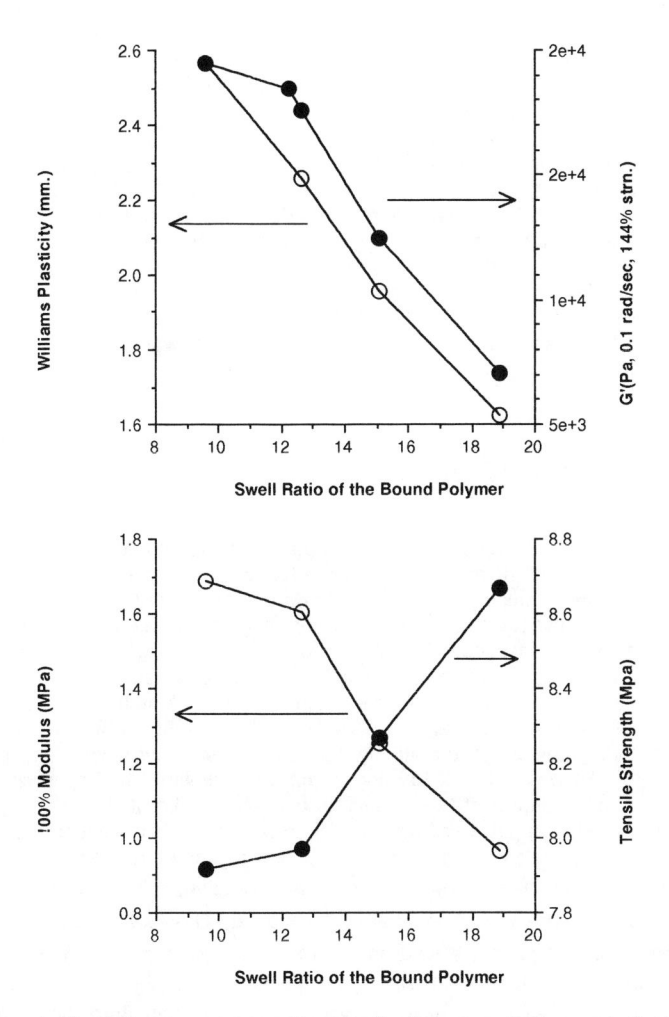

Figure 1. The influence on properties of effective cross-linking as indicated by the swell ratio in toluene of the bound polymer in PDMS elastomers that contain 30 pph of a 220 m²/g fumed silica. Upper: Properties before cure. Lower: After cure.

Unpublished work by the author [14] showed that the effective cross-linking of the bound rubber separated from an uncured PDMS elastomer filled with a high surface area fumed silica correlated with G' and Williams plasticity prior to cure. Similar trends were seen with modulus and strain at break after cure, Figure 1. In this study the content of bound polymer was essentially constant and effective cross-linking was varied through modification of the filler surface. The degree of effective cross-linking of the bound polymer was presumed to be related approximately to the inverse swell ratio of the bound polymer in a good solvent for PDMS. The results indicate that the effective cross-link density of the bound polymer influences the properties of cured and uncured PDMS. Further, in three of the four plots there appears to be a critical inflection point at a swell ratio of about 13.

The accumulated citations therefore support the existence of an extended bound polymer-filler network which appears to make major contributions to the properties of a silicone elastomer. Under strain it would contribute an additional force of recovery over that supplied by the polymer network as well as dissipate the strain energy through internal internal friction. This would take place as the network distorted, reorganized and ultimately broke down. One result would be increased resistance to premature failure and, because of the reversible nature of the adsorption of PDMS on the filler surface [15], the network would reform upon release of the strain, though not necessarily to its original configuration. Stress softening and hysteresis would be a consequence of such reorganization. A quantitative interpretation of these phenomena, which is still lacking, would offer the potential of ultimately producing a predictive theory for elastomer reinforcement.

Polymer-Filler Interfacial Energy.

A promising theoretical treatment introduced by Gent and coworkers [16] and recently refined by Edwards [17] could advance understanding of phenomena occurring at the breaking stress of elastomers reinforced with particles. The treatment was originally developed for voids in elastomer matrices and was later extended to solid inclusions. It depends on the energy required to debond the polymer from rigid spherical inclusions in an elastomer system under tension and is based on considerations of the particle diameter, the work of interfacial adhesion at the particle-matrix interface and the elasticity of the matrix. Simple dimensional considerations of the stressed matrix in the vicinity of small inclusions less than 1000 nm preclude the possibility of forming larger voids which would be precursors to failure. As the diameter of the inclusions become smaller it is increasingly more difficult to debond the polymer from the inclusion. If precursor voids formed through cavitation in the matrix are less than 100 nm in diameter, their critical applied stresses are determined by the Laplace-Young equation, $2\gamma/r$, where r is the radius of the void and γ the surface energy of the matrix. The outward applied stresses debonding the matrix from a solid inclusion will be given by $2\gamma/r_0$ where r_0 is the particle radius and γ is now the work of adhesion. Since the size of the precursor void in the vicinity of the particle surface must be smaller than the particle, the internal pressure or the equivalent triaxial tension required to sustain it will be larger than that required to debond the matrix from the particle since $r_0 > r$. This would lead to instant collapse of the void back to the surface of the particle and is consistent with the lack of void formation generally observed in stressed elastomers reinforced with small particles [17].

The reviewer applied these criteria to data from pioneering work of Warrick and Lauterbur on tensile stress at break in PDMS elastomers reinforced with particles of a variety different sizes and compositions [18]. This is illustrated in figure 2 which shows that, although there is considerable scatter, the breaking stresses fall well within the magnitude of the forces given by $2\gamma/r_0$ to debond the matrix from the polymer. Here γ was presumed to be approximately the same for the various filler interfaces. The breaking stresses can also be seen to follow roughly the predicted inverse relationship with r_0. This then suggests that the initial cavity formed on debonding filler from polymer, rather than collapsing as predicted in the previous discussion, can somehow grow to a size sufficiently large to initiate failure despite its initially small size.

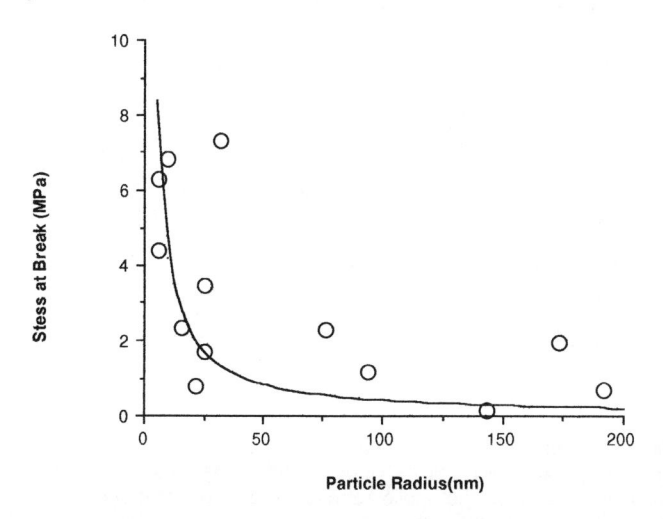

Figure 2. The effect of filler particle size on the breaking stress in a reinforced PDMS elastomer from reference 18. The particles include a variety of precipitated and pyrogenic silicas, titania, aluminum oxide and calcium carbonate. The solid line is a plot of $2\gamma/r_0$ vs r_0 where γ is presumed to be 25 mN/m.

Rigid spherical particles in isolation were assumed for reasons of simplicity in the original deliberations of Gent and Tompkins [16] to arrive at the $2\gamma/r_0$ term. However, actual dispersions of silica in PDMS are known to be much more complex and to consist of partly broken down aggregates held closely by the bound polymer in a network to give a distribution of interparticle spacings and irregular configurations. Examples of this can be seen figures 2 and 5 of reference 13. When such a network of particles is stressed to $2\gamma/r_0$, cavities would form within the confines of the surfaces of two or more proximate primary particles. The new possibility now arises that, in addition to collapsing back to the surface as originally predicted, proximate voids could fuse into one as a means of minimizing total interfacial energy per unit volume of void. Depending on the interparticle spacing and the surface energy of the polymer, the new cavity, now having a radius in the realm of r_0, could then be maintained or even grow. This is represented schematically for the simplest case in figure 3. Unlimited growth of the new void would be allowed under the applied stress when $r > r_0$. Only a single such cavity need form to initiate failure, but the probability would increase with the number of such fusions. The term $2\gamma/r_0$ would then approximate the minimum critical stress for failure. A progressive increase in resistance to failure with filler loading would then be expected until a critical separation between particles is reached where the fusion of cavities could begin. Beyond this further increases in filler would become disadvantageous as the interparticle spacing decreased. This would correspond to an optimum in resistance to failure as well as the critical volume concentration of particles at the percolation point of the polymer filler network. Examples of silicone elastomers behaving in just this fashion can be found in reference 13.

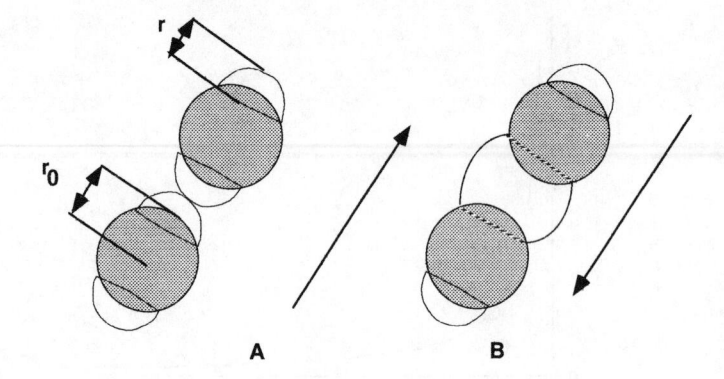

Figure 3. Schematic representation of cavitation in a stressed reinforced elastomer matrix around two proximate particles whose diameter is less than 100 nm. The arrows indicate the direction of the applied stress. A: before fusion of the cavities. B: new void formed after fusion with $r > r_0$.

RECENT APPROACHES TO REINFORCEMENT OF PDMS WITH PARTICLES

Major influence on the properties of an elastomer can be brought about through changes in the polymer, its cross-linking, changes in the reinforcement system, and in the processing where the filler is dispersed in the polymer. Because of its dramatic influence on properties, the filler has received major emphasis, especially the pyrogenic and precipitated silicas and their modifications. Until recently, however, most progress in reinforcement with particulate solids has been limited to incremental improvements over existing technologies. The approaches reviewed here were selected because of their novelty and because, if pursued, they could amount to significant technological advances in the reinforcement of elastomers in general as well as silicones.

Sol-Gel Silicas.

Particulary effective compositions for reinforcement were made by hydrolysis of mixed alkyl silicates and alkoxysilanes under carefully controlled conditions [19]. These were termed " wet process, hydrophobic silicas" or WPH by the original authors while current terminology might refer to such an approach as sol-gel processed silica. A simple rapid procedure was described where a "hydrophobing agent" such as dimethyldimethoxysilane, a silazane or a silanol was first introduced to a solution of methanol, water and ammonia. This was followed by tetramethoxysilane which gave a gel. The required amounts of reagents were defined by diagrams such as that shown in figure 4. After an aging period during which Ostwald ripening was thought to occur, the resulting gels were directly compounded as slurries into the polymer using standard equipment for mixing rubber. These gave excellent dispersions which showed minimum crepe hardening on extended storage.

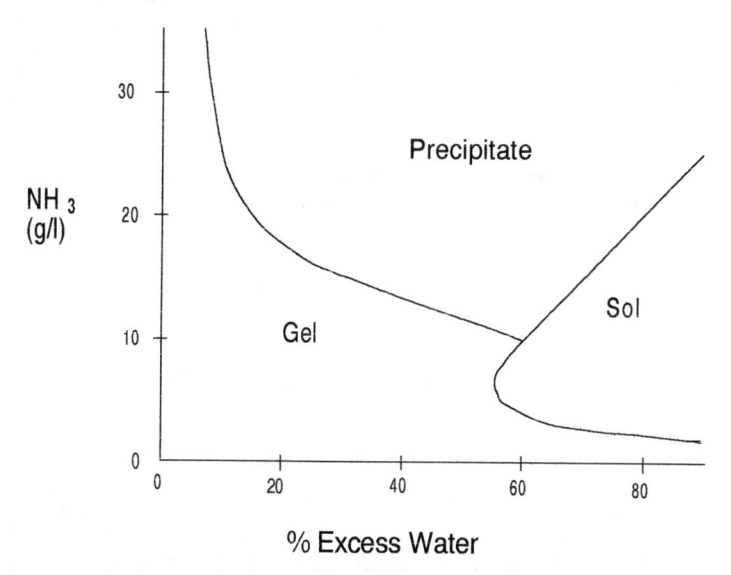

Figure 4. Example of a diagram of boundaries defining the concentrations of reagents giving sols, gels or precipitates in a WPH silica made from teramethoxysilane.

A variety of highly effective reinforcing fillers were obtained by this simple yet versatile procedure which offered the capability of designing special-purpose reinforcing fillers. This was demonstrated by preparing fillers where there was control of surface area and structure and where special surface reactivity could be introduced. An especially unique feature was optically clear dispersions in PDMS without the necessity of matching the refractive index of the filler with that of the medium. This feature combined with the excellent reinforcing characteristics offers the possibilities of some interesting new applications in the area of optics.

Although the original work was primarily with silica or closely related materials, a number of other interesting possibilities are suggested by similar sol-gel methods. Examples might be compositions based on titania, alumina, magnesia or zinc oxide or their combinations. As in the original work, an elementary understanding would be essential of the rates of hydrolysis and cohydrolysis and condensation of the alkoxides so that diagrams as that in figure 4 could be constructed.

In-situ Formation of Filler in the Elastomer Matrix.

Reinforcing particles are formed directly within the polymer or a cross-linked elastomer matrix in another novel method of reinforcing PDMS elastomers which also employs sol-gel technology [20]. This has the obvious advantage circumventing the energy consuming step of dispersing the pre-formed filler under high shear which is often accompanied by breakdown of polymer and filler. The cross-linked elastomer is first swollen with an appropriate amount of a metal alkoxide, or the alkoxide is dissolved in the polymer. The swollen elastomer, or polymer-alkoxide solution, is then exposed to water containing a catalyst which is preferably basic. The subsequent hydrolysis produces a nearly monodisperse distribution of spherical particle sizes with diameters in the range of 20-30 nm. The composition of the particles is not well understood, but particles made from ethyl orthosilicate were observed to be of much lower density than pure silica. This would suggest the presence of residual alkoxysilane structures or voids, or both. The range of particle sizes obtained are ideal for reinforcement and it was indeed shown that they provide significantly improved properties over PDMS with no reinforcing particles.

Silicone Rubber Lattices Reinforced with Aqueous Colloidal Silicas.

In another development reinforced silicone elastomers were obtained from lattices formed by simply mixing aqueous colloidal dispersions of silica with aqueous anionically stabilized emulsions of high molecular weight PDMS. The mixed colloids were then aged under alkaline conditions in the presence of tin catalysts commonly used for silanol condensations [21]. The PDMS emulsion particles were found to cross-link during the aging process, probably through condensation of the silanol chain ends of the PDMS with the soluble silicates present in the colloidal silica. The final result is a stabile dispersion of cross-linked PDMS particles 100 - 500 nm in diameter and silica particles 5 - 10 nm in diameter. Removal of the water through evaporation, leaves elastomeric films where, according to electron microscopy, the cross-linked PDMS particles are encapsulated by the colloidal silica particles. This is represented in the upper part of figure 5 and is just the reverse of a typical elastomer where the particulate solid is normally dispersed within the cross-linked polymer.

When placed in tension the films exhibited a relatively high Young's modulus and yield points whose magnitude depended on the amount of colloidal silica. At strains beyond the yield point modulus was more that of a typical silicone elastomer. It was suggested that the structure represented in the upper part of figure 5 dominated prior to the yield point at low strain, while after the yield point the nearly continuous silica structure was destroyed as the strain increased. After the yield point the structure represented in the lower part of figure 5, which approached more a dispersion of particles within PDMS rather than the reverse, was considered more dominant.

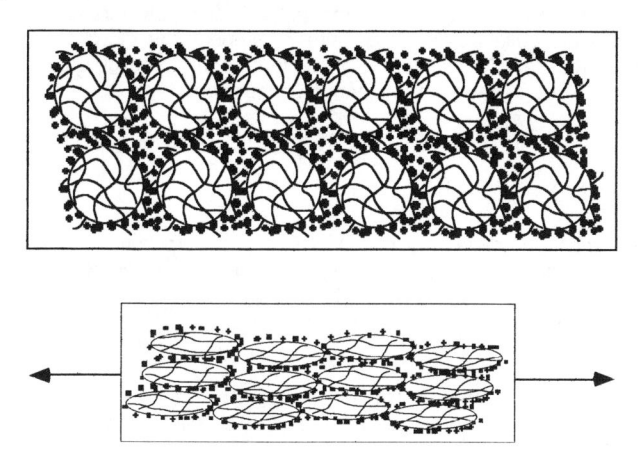

Figure 5. Schematic representation of a silicone rubber latex film after evaporation of the water (upper) and in tension (lower).

The silicone latex system and the use of an aqueous colloid for reinforcement provides yet another simple alternative to introducing the filler to a high molecular weight PDMS without resorting to high shear mixing equippment. It also offers the special advantages of aqueous emulsion polymers, such as low viscosity dispersions of high polymers and a dispersion medium compatible with the environment.

FUTURE PROSPECTS

Although there are many rewarding areas for future research and development in reinforcement of silicone elastomers with particulate solids, the most significant progress will be in developing a coherent theory for reinforcement. The importance of the filler-polymer network and the surface tension effects must both be taken into account. In addition the capability of at least approximate quantitative predictions of elastomer properties should be a requisite of any theory of reinforcement. The most promise might be offered by an approach utilizing percolation theory to describe the polymer-filler network while taking into consideration the effect of interfacial energy of adhesion of the polymer to the filler as reviewed above.

With sound theoretical background as a guide it should be possible to design or "engineer" reinforcement systems for PDMS by taking advantage of the recent advances in colloid science that can provide particles in a variety of compositions sizes, shapes and configurations. This would also require new approaches for dispersing the reinforcing filler in the polymer matrix so that the carefully designed particle structures would not be destroyed by the time worn methods of the high shear mixing process now commonly in use for the manufacture of elastomers. This might include polymerization around the filler particle, forming the filler particle in the polymer matrix, dispersing the polymer and filler in solvents such as supercritical fluids and simply mixing aqueous colloids and emulsion polymers. The approaches reviewed here are believed to be first faltering steps in this direction and they may well be the forbearers for the future of reinforcement. The guidance of a sound theoretical background should carry them much further.

ACKNOWLEDGEMENT

The author acknowledges the Dow Corning Corporation for its support in preparing this manuscript and for granting permission to review the data in figure 1.

REFERENCES

1. J. E. Mark, Silicon Polymer Based Science, Adv. in Chem. Ser., J. M. Zeigler and F. W. G. Fearon Eds., Amer. Chem. Soc., Div. of Polym. Chem., 1990, pp. 47. M. J. Owen, Ibid., pp. 705.

2. E. L. Warrick, O. R. Pierce, K. E. Polmanteer, and J. C. Saam, Rubber Chem. Technol. 52 (3), 437, (1979).

3. T. L. Smith, Polym. Eng. Sci., 17, 129, (1977).

4. J. E. Mark and M. Y. Tang, J. Polym. Sci., Polym. Phys. Ed., 22, 1849, (1984).

5. T.L. Smith, B. Haidar, and J. L. Hedrick, Rubber Chem. and Technol., 63, 256, (1990).

6. B. B. Boonstra, H. Cochrane, and E. M. Dannenberg, Rubber Chem. and Technol., 48, 558, (1975).

7. B. B. Boonstra, Polymer, 20, 691, (1979).

8. E. Dannenberg, Rubber Chem. and Technol., 48, 410, (1975).

9. D. W. Southwart and T. Hunt, J. Inst. Rubber Ind., 3, 249, (1969). Ibid., 4, 74, (1970).

10. A. V. Galanti and L. H. Sperling, J. Appl. Polym. Sci., 14, 2785, (1970); Polym. Eng. Sci., 10, 177, (1970).

11. M. Arenguren, Thesis, University of Minnesota, (1990).

12. A. R. Payne, R. E. Whittiker and J. F. Smith, J. Appl. Polym. Sci., 16, 1191, (1972).

13. A. Pouchelon and P. Vondracek, Rubber Chem and Technol., 62,788, (1989).

14. J. C. Saam and F. R. Minger, unpublished.

15. G. Berrod, A. Vidal, E. Papirer and J. B. Donnet, J. Appl. Polym. Sci., 23, 2579, (1979).

16. A. N. Gent and D. A. Thompkins, Rubber Chem . and Technol., 43, 873, (1970). A. N. Gent, Ibid., 63(3), G49, (1990). A. N. Gent , J. Mater. Sci., 15, 2884, (1980).

17. D.C. Edwards, J. Mater. Sci., 25, 4175, (1990).

18. E. L. Warrick and P. C. Lauterbur, Ind. Eng. Chem., 47, 486, (1955).

19. M. A. Lutz, K. E. Polmanteer and H. L. Chapman, Rubber Chem. and Technol., 58, 939, (1985); 58, 953, (1985); 58, 965, (1985).

20. For a more detailed review and leading references see J. E. Mark, H. R. Allcock and R. W. West, Inorganic Polymers, Prentice Polymer Science and Engineering Series, edited by J. E. Mark, Prentice Hall, Englewood Cliffs, New Jersey, 1992, pp. 173-185.

21. J. C. Saam, D. Graiver and M. Baile, Rubber Chem and Technol., 54, 976, (1981).

INTERPENETRATING ORGANOMETALLIC POLYMER NETWORKS (IOPN's): NOVEL ADVANCED MATERIALS

B. CORAIN[*,¤], M. ZECCA[*], C. CORVAJA[*], G. PALMA[*], S. LORA[¤] and K. JERABEK[◊]
[*]University of Padova, Dept. Inorganic Chemistry, via Marzolo 1, and Dept. Physical Chemistry, via Loredan 2, I35131 Padova, Italy;
[¤]C.N.R.,Centro Studio Stabilità e Reattività Composti Coordinazione, via Marzolo 1, I35131 Padova, Italy and Institute F.R.A.E., I35020 Legnaro, Italy;
[◊]Czechoslovak Academy of Science, Institute Chemical Processes Fundamentals, 165 02 Praha 6 - Suchdol, Czechoslovakia.

ABSTRACT

A cross-linked copolymer of vinylferrocene and dimethyl-acrylamide is generated inside the less dense microporous domains of a cross-linked polydimethylacrylamide matrix. I.S.E.C. and E.S.R. data reveal that the obtained composite is in fact an I.P.N.(Interpenetrating Polymer Network) material.

INTRODUCTION

The dispersion of metal centres inside inorganic or organic matrixes is a key-procedure for obtaining supported metal catalysts [1,2] or specialties such as materials for recording devices, antifouling paints, electrodes modifying coatings, etc. [3].

The main technological procedures currently used for dispersing metal centres inside high surface-area supports are i) impregnation, ii) ion exchange and iii) deposition-precipitation of metal ions initially dissolved in convenient solutions [2]. These technological procedures account for most of the presently employed industrial catalysts [4] and normally the role of the support is played by inorganic oxides.

The possibility of utilizing organic, thermally stable, designable, mechanically robust supports became apparent to the scientific community in the years following the publication of three seminal patents by Haag and Whitehurst in 1969 [5]. The original proposal of these authors was mainly directed to the development of "hybrid catalysts" [6], i.e. materials able to combine the best features of both homogeneous and heterogeneous catalysts, with the obvious expectation of eliminating their specific disadvantages [7]. This target proved to be unattainable, so far, in real industrial terms [8], but the huge scientific production induced by this "race to the perfect catalyst" [7,9] has led to a fairly diffuse awareness that organic matrixes could be most valuable supports [10,11,12] for obtaining metal catalysts potentially useful for fine-chemicals synthetic technologies.

The switch point in this connection is the realization that a properly designable cross-linked matrix may be a most convenient support [12] for dispersing zero valent metal centres in the form of very fine (nanometers-sized) metal particle, upon starting from solid, totally isotropic macromolecular metal complexes [13].

RESULTS

Possible synthetic strategies for preparing *ad hoc* macromolecular metal complexes are depicted in Figure 1.

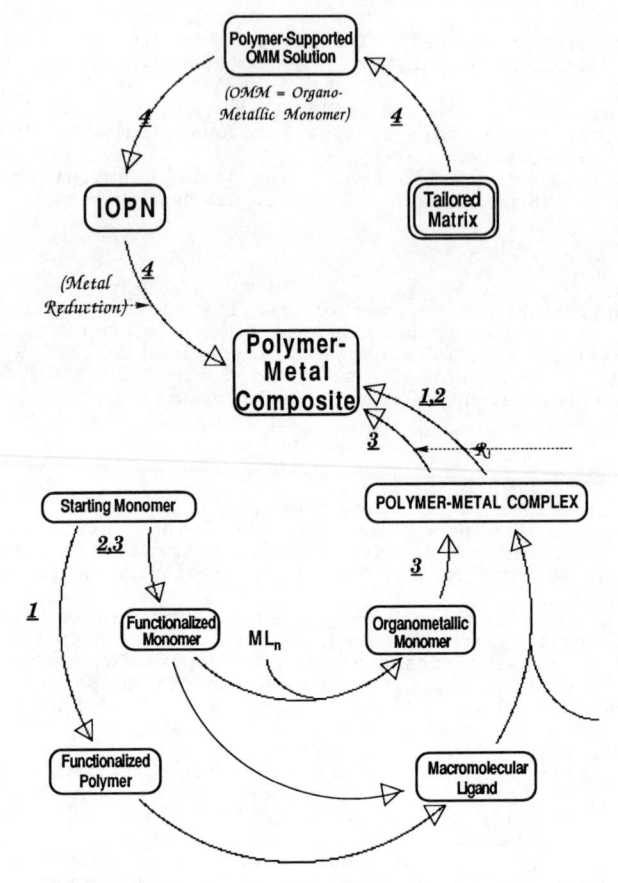

Figure 1. Major synthetic procedures for the homogeneous dispersion of metal centres throughout designed synthetic matrixes.

Route **1** appears to be by far the most exploited one in the relevant literature [7]. Route **2** is comparatively almost ignored [14]. Route **3** has been thoroughly investigated mainly by russian scientists [15], but with little apparent consideration for chain topology and solvent-substrate compatibility [12], i.e. for matrix "molecular engineering".
Route **4** is a proposal stemming from these laboratories [16,17]. A convenient matrix (polydimethylacrylamide cross-linked with methylenebisacrylamide, 4% mol/mol) is prepared and swollen with a solution containing a) an organometallic or metallo-organic

(vide infra) comonomer, b) a properly chosen auxiliary comonomer, c) a suitable cross-linking agent, d) if required, a convenient polymerization catalyst. The solvent must be carefully selected in that it has to combine (co)monomers solubilization ability and matrix swelling capacity. After soaking the matrix with just the required amount of reacting solution, the copolimerization can be performed, either thermally, or catalytically or upon exposure to γ-rays [18]. If the guest copolymer undergoes an intimate (i.e. molecular) dispersion inside the pre-existing polymer network of the supporting host matrix, the final material will be considered an I.P.N. (Interpenetrating Polymer Network) system [19]. We developed the strategy to organometallic I.P.N's (route **4**) by utilizing vinylferrocene as paradigmatic organometallic comonomer, dimethylacrylamide as auxiliary comonomer and methylenebisacrylamide as cross-linking agent (4 % mol/mol) [16]. The combination of elemental analyses, IR and Mössbauer spectrometries, SEM and X-ray microprobe analyses led us to a 'first level' satisfactory characterization of the organometallic composites obtained along route **4**.

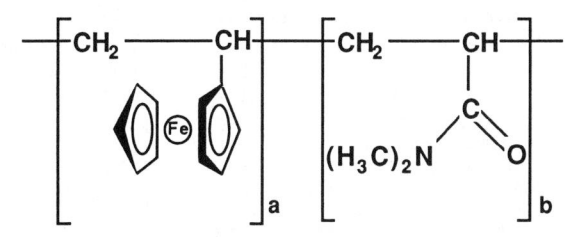

Figure 2. Sketch of the molecular structure of the guest organometallic copolymer.

The *structural characterization* of these materials, viewed both in terms of a) *chain density modifications* as the consequence of switching from the matrix to the expected I.P.N. material, b) *molecular accessibility* of their microporous domains and c) *chemical accessibility* of the metal centres decorating the surfaces of the nanometer-sized cavities, represented an appropriate 'second level' analysis of these materials.
I.S.E.C. (Inverse Steric Exclusion Chromatography) [20], **E.S.R.** spectrometry of selected spin label dissolved in solutions soaking the composites [21,22] and **chemical oxidation** of the ferrocenyl moieties by selected agents [23] proved to be useful tools for this evaluation [16,17].

I.S.E.C. Analysis.

The matrix (**I**) and the composite (**I-Fe**) were used as stationary phases in chromatographic measurements with water, dichloromethane (DCM) or tetrahydrofurane (THF) alternatively, as mobile phase. From the elution behavior of solutes with known molecular size (D$_2$O, sugars and dextranes in water, alkanes and polystyrenes in DCM and THF), important information on swollen polymer mass density distribution in the studied materials was obtained [17]. The results of I.S.E.C. analyses on **I** and **I-Fe** are collected in Table I. In **I** swollen in water or DCM, the

Table I. Results of I.S.E.C. analysis [a] on **I** and **I-Fe** [ref. 17]

Polymer Chain Density (nm/nm^3)	Fraction Volume per unit mass of dry material (cm^3/g)	
	I	I-Fe
0.1	0.000	0.074
0.2	0.930	0.108
0.4	0.212	0.068
2.0	1.340	1.603

(a) water as mobile phase; for a full description of this technique see ref. 20 and refs. therein.

investigation reveals beside dense backbone (chain density ca. 2 nm/nm^3) the presence of a substantial volume of low-density domains (chain density < 1 nm/nm^3). The I.S.E.C. analysis of **I-Fe** shows the presence of the higher density polymer mass only. This is reflected in the volume-averaged polymer chain density ($\bar{c}$) of the gel regions which can be readily calculated for both materials from the data of table I. $\bar{c}$ values of 1.19 nm/nm^3 and 1.76 nm/nm^3 are found for **I** and **I-Fe**, respectively. The comparison of these two figures can be interpreted in terms of reduced "porosity" of the composite with respect to the parent matrix. Thus, the composite appears to be in fact an I.P.N. material in which the low-density domains seen in its precursor **I** have disappeared as they have become preferential hosting volumes for the accomodation of the guest organometallic copolymer.

E.S.R. Analysis.

E.S.R. study of motion of paramagnetic probes is a recognized tool for the evaluation of the steric hindrance inside cavities of "molecular" size, like e.g.intrazeolitic micropores [21]. The molecular accessibility of the interior of swollen polymer domains should be similarly evaluated by comparing the rotational mobility of a suitable probe (2,2,6,6-tetramethyl-4-oxo-1-oxyl-pyperidine, TEMPONE) dissolved either in a free solution or in a solution confined in the swollen polymeric gel. Our preliminary data reveal that the rotational mobility at room temperature of the probe in water (τ = 54 ps,[24]) decreases quite markedly when aqueous solutions of TEMPONE are employed to swell both **I** and **I-Fe** (τ = 220 and 520 ps, respectively). These data show that the comparatively dense gel regions still present in the organometallic I.P.N. are accessible to molecules with substantial sizes, such as TEMPONE (ca. 0.2 nm^3). On the other hand, high $\bar{c}$ values are expected to considerably reduce [25] the rotational mobility of the paramagnetic probe. This is fully confirmed by our E.S.R. data which are in excellent agreement with the I.S.E.C. characterization of the morphology of **I** and **I-Fe** [17]. A good linear fit between log(τ) and $\bar{c}$ is found, but further experimental data are needed for confirmation of this linear relationship.

$$\lg(\tau) = 52.6 \times 10^{(0.552\ \bar{c})}$$
$$r = 1.00$$

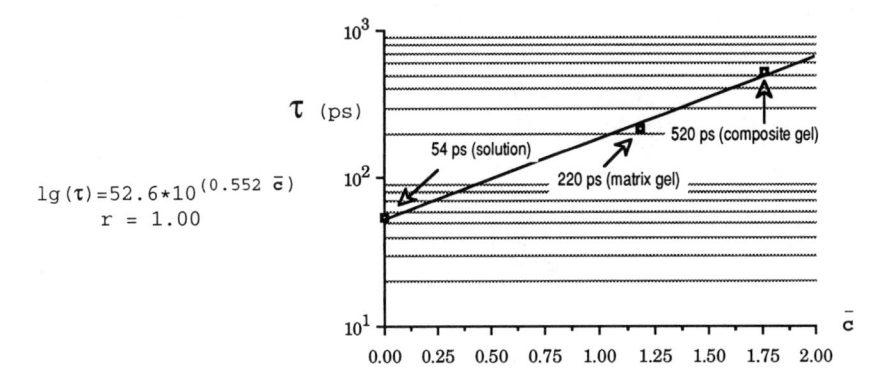

Figure 3. Rotational correlation time of TEMPONE in water vs $\bar{c}$ in the investigated systems

Chemical Reactivity.

The ferrocenyl moieties in **I-Fe** are readily oxidized to ferricinyl ones by equimolar amounts of Ag$^+$, upon addition of proper amounts of 0.25 M aqueous solutions of silver nitrate to the water-swollen I.O.P.N. Quantitative oxidation is achieved after 22 hours, as shown by quantitative elemental analysis of metal silver formed upon Fe^{II} oxidation and concomitantly deposited throughout the polymer network. Extensive Fe^{II} oxidation, i. e. 63 and 87 % is observed after 0.5 and 3.0 hours respectively.

CONCLUSIONS

Organometallic moieties can be evenly distributed inside preformed tailored matrixes *via* a copolymerization procedure in solvents able to swell the hosting matrix. The appropriate choice of cross-linking degree for both the host and the guest components and of the experimental conditions appears to make possible the "molecular" dispersion of the guest copolymer and to provide novel materials possessing remarkable molecular accessibility and chemical reactivity.

ACKNOWLEDGEMENTS.

This work was partially supported by Progetto Finalizzato Chimica Fine II, C.N.R. Rome and by Ministero dell'Università e della Ricerca e Tecnologia, Rome, fondi 40%.

REFERENCES.

1. K. Foger in *Catalysis*, edited by J.R. Anderson and M. Boudart (Springer Verlag, Berlin, 1984).
2. G.C. Bond, Chem. Soc. Rev., 20, 441 (1991).
3. a) C. U. Pittman Jr. and M. D. Rausch, Pure Appl. Chem., 58, 614 (1986). b) J.E. Sheats in *Kirk-Othmer "Encyclopedia of Chemical Technology"*, 3rd ed. (John Wiley & Sons, Inc., New York, 1981), vol. 15, pag. 184.
4. a) R. Pearce and W.R. Patterson, *Catalysis and Chemical Processes* (Blackie & Son Ltd., Glasgow, 1981). b) C.N. Satterfield, *Heterogeneous Catalysis in Practice* (McGraw-Hill Book Company, New York, 1980).
5. W.O. Haag and D.D. Whitehurst, German Patents 1800371 (1989), 1800379 (1969), 1800380 (1969).
6. G.W. Parshall, *Homogeneous Catalysis* (Wiley-Interscience Publication, New York, 1980), p. 227.
7. F.R. Hartley, *Supported Metal Complexes* (D. Reidel, Dordrecht, 1985).
8. See ref 7, p. 3.
9. C.U. Pittmann Jr., in *Comprehensive Organometallic Chemistry*, edited by G. Wilkinson et al. (Pergamon, Oxford, 1982) Vol. 8, p. 553.
10. J. Lieto, D. Milstein, R. L. Albright, J. V. Minkiewicz and B. C. Gates, Chemtech, 1983, 46.
11. B.C. Gates, *Catalytic Chemistry*, (John Wiley & Sons, Inc., New York, 1992), p. 182.
12. R. Arshady, Adv. Mater., 3, 182 (1991).
13. For a recent example see B. Corain, M. Basato. M. Zecca, G. Braca, A.M. Raspolli Galletti, S. Lora, G. Palma and E. Guglielminotti, J. Mol. Catal., 73, 23 (1992).
14. B.Corain, F.O. Sam, M. Zecca, A.C. Veronese, G. Palma and S. Lora, Makromol.Chem.Rapid Commun., 10, 69 (1990).
15. A.D. Pomogailo and V.S. Savastyanov, IMS-REV Macromol. Chem. Phys., C25, 375 (1985).
16. R. Arshady, B. Corain, S. Lora, G. Palma, U. Russo, F.O. Sam and M. Zecca, Adv. Mater., 2, 412 (1990)
17. B. Corain, K. Jerabek, S. Lora, G. Palma and M. Zecca, Adv. Mater., 4, 97 (1992).
18. R. Arshady, M. Basato, B. Corain, L. Della Giustina, S. Lora, G. Palma, M. Roncato and M. Zecca, J. Mol. Catal., 53, 111 (1989) and references therein.
19. L. H. Sperling, Chemtech, 104 (1988).
20. K. Jerabek, Anal. Chem., 57, 1598 (1985).
21. G. Martini, M. F. Ottaviani and M. Romanelli, J. Colloid Interface Sci., 115, 87 (1987).
22. R. D. Harvey and S. Schlick, Polymer, 30, 11 (1989).
23. F. A. Cotton and G. Wilkinson, *Advanced Inorganic Chemistry*, 5th ed. (John Wiley & Sons, Inc., New York, 1988), p. 1173.
24. Various authors in *Spin labelling Theory and Application*, edited by L.J. Berliner (Academic Press, New York, 1976).
25. G. Nimtz, in *Correlations and Connectivity*, edited by H.E. Stanley and N. Ostrowsky (NATO ASI Series, 118, Kluwer, Dordrecht, 1990), p. 225.

SYNTHESIS, CHARACTERIZATION, AND DYNAMICS OF A ROD/SPHERE ORGANOCERAMIC COMPOSITE LIQUID

MARK A. TRACY* AND R. PECORA*
*Stanford University, Department of Chemistry, Stanford, CA 94305-5080

INTRODUCTION

Composite liquids – liquids composed of polymers, particles, and small molecule solvents – constitute an important class of synthetic and naturally occurring materials. Examples include molecular composites, ceramic precursors, lubricants, adhesives, and the cytoplasm in biological cells. Due to the complexity of these liquids, experimental studies of precisely defined systems are essential in developing an understanding of the interactions between all components in the liquid. Unfortunately, such fundamental studies have been relatively rare due to both the difficulty of synthesizing precisely defined composite liquids and the lack of adequate experimental methods to monitor the motions of the various constituents.

We have recently reported the synthesis, characterization and some studies of the dynamics of a rod/sphere composite liquid system [1]. In our case the "polymer" constituent is a rigid rod polymer, poly(γ-benzyl-α,L-glutamate) (PBLG). Rigid rod polymers are frequently used in composite liquids as viscosity enhancers. PBLG is commercially available in a wide range of molecular weights and its static and dynamic behavior in dilute and nondilute solutions has been studied. It, in addition, forms mesophases in the concentrated regime. The ceramic "particles" in our composite liquid are coated silica spheres. These spheres are synthesized by the method of Stöber et. al. [2] and coated with an organic coating (3-(trimethoxysilyl)propyl methacrylate (TPM)) following a procedure based on that of Philipse et. al. [3] to render them dispersible in organic solvents. The spheres with sizes in the range from 10 nm up to almost 1µm can be synthesized with a relatively narrow size distribution. The solvent in our studies is dimethylformamide (DMF). Both polymer and particle are dispersible as singlets (non-aggregating) in these solvents and the PBLG retains its rigid (or nearly rigid) rod conformation. The diffusion of both the polymer and the sphere in the composite liquid is measured by dynamic light scattering (DLS) [4]. In this paper, we focus on the spheres and examine the effects of rod concentration and rod length on the diffusion of different size spheres. This study suggests that the solution microstructure has an important influence on sphere diffusion.

EXPERIMENTAL

Coated silica spheres of radii 39.4 nm and 60.4 nm (determined by DLS) were synthesized by methods described previously [1]. PBLG of molecular weights 102,000 and 249,700 g/mol were purchased from the Sigma Chemical Company.

The dynamics of both binary rod/DMF and sphere/DMF solutions were studied by DLS in addition to the rod/sphere/DMF composite liquid. The rod concentrations studied varied from the dilute to the semidilute regime where rod overlap becomes important [1]. The sphere concentrations were dilute (0.047-0.28 mg/ml (0.005-0.03%)). Sphere/DMF solutions were studied but showed no change in the diffusion constant over the dilute concentration range present in these experiments. DLS data were taken at a minimum of three of the following angles: 25.45° or 30.56°, 59.44°, 90°, and 110.35° or 120.56°. Results are reported here for three composite liquids: Composite Liquid 1 (CL1) consists of 60.4 nm spheres and PBLG of molecular weight 102,000 g/mol in the concentration range of 1-30 mg/ml ($2<nL^3<60$, where n is the rod number concentration and L is the rod length (70 nm)). Composite Liquid 2 (CL2) consists of 60.4 nm spheres and 249,700 g/mol PBLG in the concentration range 1 to 13.1 mg/ml ($12<nL^3<158$) Finally, Composite Liquid 3 (CL3) is a mixture of 39.4 nm spheres and 249,700 g/mol PBLG in the same concentration range ($12<nL^3<158$). All experiments were done at 25°C. Solution viscosities were measured with an Ostwald capillary viscometer at 25°C.

The intensity autocorrelation function is measured in the homodyne DLS experiments using a standard DLS apparatus [1]. For the solutions described here, the diffusion constants of the rods and the spheres were determined from the experimentally determined correlation function using the FORTRAN program CONTIN which calculates the distribution of decay times (τ_R) leading to the exponential decay of the measured correlation function. The decay time is inversely proportional to the diffusion constant, D:

$$\tau_R = 1/q^2 D \qquad (1)$$

where q is the magnitude of the scattering vector,

$$q = (4\pi n/\lambda_0)\sin(\theta/2) \qquad (2)$$

and n is the solvent index of refraction, λ_0 the wavelength of laser light in a vacuum, and θ the scattering angle. In the CONTIN results shown in Figure 1, the distribution of decay times is presented as a plot of relative intensity versus hydrodynamic radius (R_h) of the scattering species. R_h is calculated from D using the Stokes-Einstein equation:

$$D = kT/6\pi\eta R_h \qquad (3)$$

where k is the Boltzmann constant, T the absolute temperature, and η the solution viscosity. The solvent viscosity of DMF was used in place of the solution viscosity in the outputs to provide a standard for comparison of all peaks. The intensity at a given R_h is the fraction of scattered light from particles of "size" R_h.

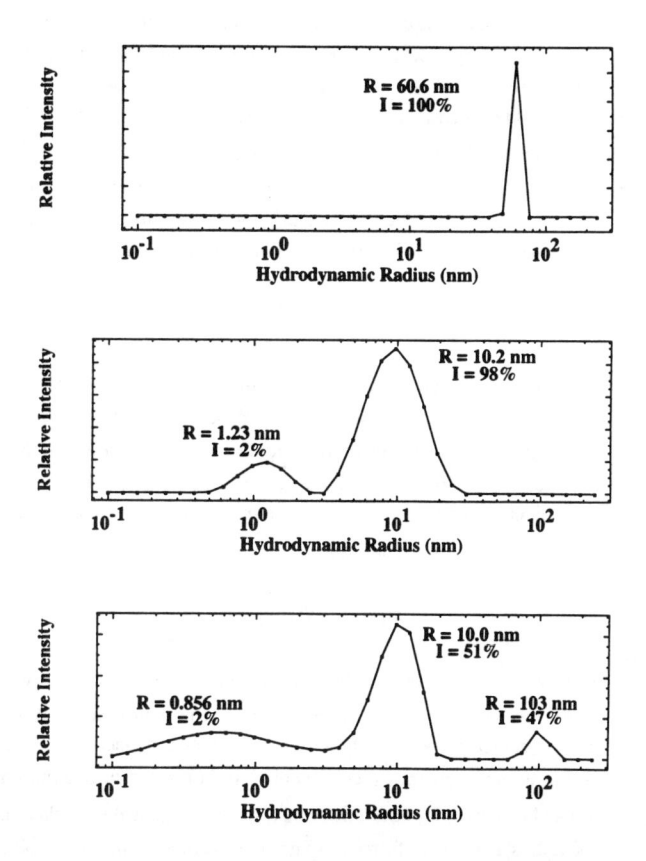

Figure 1. Comparison of binary solution DLS results to those of the ternary composite liquid. The top figure is a typical intensity versus hydrodynamic radius plot obtained from CONTIN for a 0.01% silica/DMF solution. Each CONTIN peak is marked by an R and an I. R is the average hydrodynamic radius of the peak and I is the fraction of the total scattered intensity within that peak. The second plot is a typical CONTIN output for a 5.0 mg/ml PBLG/DMF solution. The bottom plot is a typical output for the composite liquid showing both the rod and the sphere diffusion. This composite liquid contained 5.0 mg/ml rods and 0.005% spheres in DMF. For each output, the solvent viscosity, not the solution viscosity, was used to calculate R. Thus R for the sphere varies from the value in the sphere/DMF output. These examples were taken from data measured at 90°.

RESULTS

As shown in Figure 1, both the rod and the sphere diffusion constants were measured simultaneously in the composite liquid solutions. Figure 1 also shows that the peaks in the composite liquid DLS results were unambiguously identified from a comparison with the binary sphere/DMF and rod/DMF DLS results. In all the composite liquid data analyzed, two dominant peaks were resolved by CONTIN. The slow peak was attributed to the translational diffusion of the spheres because as the polymer concentration decreased, the hydrodynamic radius approached that of the spheres alone in DMF. The faster peak (smaller R_h) is due to the translational diffusion of the rods. Both the sphere and rod diffusion constants are independent of angle. The very fast, weak third peak is most likely due to the translational-rotational coupling of the rod diffusion [5]. The effect of increasing rod concentration on the rod diffusion constant is discussed in detail in [1]. Here we focus on the sphere diffusion.

DISCUSSION

The spheres in these composite liquid solutions are utilized as probes of the composite solution structure. We are interested in addressing the question: Is the solution microstructure relevant in determining sphere diffusion? Information about the microscopic solution environment encountered by the spheres as they diffuse can be obtained by using the Stokes-Einstein equation to calculate the local viscosity (or microviscosity, η_μ) from the measured diffusion constants for spheres of known radius R:

$$\eta_\mu = kT/6\pi RD \qquad (4)$$

If the microviscosity is identical to the solution viscosity (η), measured here by a capillary viscometer, the solution behaves as a structureless continuum-a medium completely characterized by a macroscopic property of the solution: the solution viscosity η. However, if $\eta_\mu < \eta$, the local solution environment (microstructure) is important in determining sphere diffusion. Langevin and Rondelez [6] have presented a scaling argument to explain this case in which the probe of radius R diffuses faster through a semidilute solution of polymers than predicted from the macroscopic viscosity using Equation 3. Their argument is based on topological considerations which treat the background polymer above the polymer overlap concentration c* as a transient "net". Their argument leads to the following equation for the microviscosity:

$$\eta 0/\eta_\mu = \eta 0/\eta + \exp(-R/\xi) \qquad (5)$$

where 0 signifies zero polymer concentration and ξ is the "correlation length" between polymer molecules. ξ is a measure of the mesh size of the net. The mesh size can be estimated as the average distance between the rods:

$$\xi = (M/cN_a)^{1/3} \qquad (6)$$

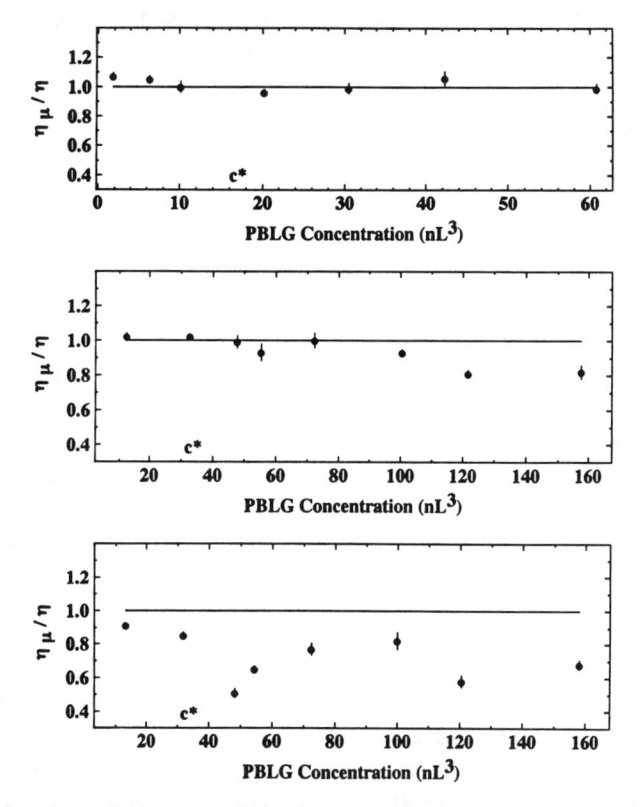

Figure 2. Microviscosity/Solution viscosity (η_μ/η) for the three composite liquids studied as a function of the rod concentration given in dimensionless units (nL^3). The solid circles are the microviscosities calculated using Equation 4. Figure 2a (top) shows results for CL1, Figure 2b (middle) for CL2, and Figure 2c (bottom) for CL3. c* is the polymer overlap concentration determined as described in [1].

According to Equation 5, if $R>>\xi$, $\eta_\mu \approx \eta$. But, as the probe size becomes less than the net hole size, the probes are able to diffuse through the holes and, thus, encounter a local viscosity that approaches that of the solvent, DMF. As the rod concentration increases, the mesh size decreases so η_μ again approaches η. The situation $\eta_\mu>>\eta$ is often indicative of aggregation.

These ideas were tested by systematically varying the sphere size, rod length, and rod concentration. The results for CL1-3 are shown in Figure 2. For CL1, $2.4<R/\xi<3.4$ above the rod overlap concentration. Similarly, $1.1<R/\xi<1.9$ for CL2 and $0.7<R/\xi<1.2$ for CL3. Figure 2a shows that $\eta_\mu \approx \eta$ at all concentrations for CL1. Figures 2b and 2c show that at the low rod concentrations, $\eta_\mu \approx \eta$ for CL2, but η_μ was up to a factor of two smaller than η for CL3. The only difference between CL2 and CL3 is the sphere size. The deviation from $\eta_\mu/\eta=1$ is largest in Figure 2c near the concentration at which the rods begin to overlap (c*) [1] and then decreases. This is qualitatively in accordance with Equation 5 where the ratio R/ξ must be sufficiently small in order to measure $\eta_\mu < \eta$. This qualitative agreement with equation 5 suggests that above c* the rod solution can be treated as a net as far as the diffusion of the spheres is concerned. The sphere is sufficiently small so that it diffuses through holes in the net. Adding more polymer decreases the hole size. Thus, gaps in the net become too small to be probed by the spheres and η_μ again approaches η.

Deviations from $\eta_\mu/\eta=1$ at the highest rod concentrations were found for both CL2 and CL3. Similar behavior has been observed before by other investigators [7-10] studying sphere diffusion in solutions of coil polymers at high polymer concentrations. Although no theory exists to explain this behavior, it has been attributed to coupling between the rod and sphere motions at the higher concentrations [9-10]. It may also be due to a transition in the solution structure brought on by presence of the spheres. For example, polymers have been found to induce a phase separation in colloidal suspensions [11-12]. A clear solution turning cloudy is usually indicative of such a transition. This was not observed in our samples. At high rod concentrations, the rods are known to form ordered (liquid crystalline) regions [13]. The influence of the spheres on the ordering tendencies of the rods may be reflected in the results at high concentration in Figures 2b and 2c. Studies are now underway to investigate this question.

CONCLUSION

We have synthesized a new rod/sphere composite liquid and performed DLS experiments to study the rod and sphere dynamics at a variety of rod concentrations. The diffusion constants of both the rods and the spheres were measured simultaneously using DLS. The local viscosity encountered by the spheres was found to be up to a factor of two times smaller than the solution viscosity. This depended upon the ratio of the sphere radius to the average mesh size of the rods in qualitative agreement with the picture of Langevin et. al. [6] in which the polymer solution is viewed as a"net" with holes rather than as a continuum. Thus, solution topology is important in determining sphere diffusion in this composite liquid.

REFERENCES

1. M.A. Tracy and R. Pecora, Macromolecules 25, 337 (1992).
2. W. Stöber, A. Fink, and E. Bohn, J. Colloid Interface Sci. 26, 62 (1968).
3. A. Philipse and A. Vrij, J. Colloid Interface Sci. 128, 121 (1989).
4. B.J. Berne and R. Pecora, Dynamic Light Scattering, (John Wiley, New York, 1976).
5. K.M. Zero and R. Pecora, Macromolecules 15, 87 (1982).
6. D. Langevin and F. Rondelez, Polymer 19, 875 (1978).
7. G.D.J. Phillies, J. Phys. Chem. 93, 5029 (1989).
8. T. Yang and A.M. Jamieson, J. Colloid Interface Sci. 126, 220 (1988).
9. P. Zhou and W. Brown, Macromolecules 22, 890 (1989).
10. W. Brown and R. Rymdén, Macromolecules 21, 840 (1988).
11. A.P. Gast, C.K. Hall, and W.B. Russell, J. Colloid Interface Sci. 96, 251 (1983).
12. P.R. Sperry, J. Colloid Interface Sci. 99, 97 (1984).
13. L.M. DeLong and P.S. Russo, Macromolecules 24, 6139 (1991).

SILICA/SILICONE NANOCOMPOSITE FILMS: A NEW CONCEPT IN CORROSION PROTECTION

Theresa E. Gentle and Ronald H. Baney
Dow Corning Corp., Midland, MI. 48686

ABSTRACT

Thin films of silsesquioxane, $(HSiO_{3/2})_n$, were applied to aluminum panels and to CMOS microelectronic circuit surfaces by spin or dip coating organic solutions of the silsesquioxane. Nanoporous silica was obtained by oxidation of the silsesquioxane. These nanoporous silica films were then vacuum infiltrated with various viscosities of polydimethylsiloxanes (PDMS) to form hydrophobic nanocomposites. The nanocomposite films were shown to provide superior hermetic protection against salt fog exposure when compared to PDMS and silica films alone.

The composite films were characterized by FTIR and optical microscopy. FTIR spectra showed that the silica served as a skeletal framework holding the hydrophobic PDMS in place and preventing loss of adhesion. This is in contrast to PDMS films alone in which blistering of the film from the substrate can occur, thus, allowing ions and moisture to reach the surface and corrosion to take place.

INTRODUCTION

Aluminum corrosion requires three conditions to occur [1]. There needs to be liquid water at the surface of the aluminum, oxygen transport to the surface and chloride ions at the surface to break up the naturally occurring alumina passivation layer. Chloride ion transport to the surface of the aluminum would most likely require liquid water. The amount of water, chloride ion, and oxygen need not be great for corrosion to occur. However, if the transport of these components can be prevented or at least reduced, corrosion can be retarded.

Silicones have commonly been used in corrosion protection of electronic devices. Protection by silicone gels is reported to fail by blistering from the substrate with water bubbles forming at the interface of the aluminum and the silicone [2, 3]. Ions at the circuit surface promote water condensation and the resulting osmotic pressure causes the blistering.

Some corrosion protection can be offered by coating with silica. However, depending on how the silica is derived, it can be quite porous which allows the passage of ions and moisture to the substrate surface. For instance, sol-gel derived silica is quite porous unless a high temperature densification step is used. Further, the high temperature densification can lead to cracking of the silica which can also provide a pathway for ions and moisture to reach the substrate.

Haluska et al. [4, 5] have described a new silica precursor that can be applied from solution by spin, drop, dip, or flow coating techniques. This silica precursor, hydrogen silsesquioxane, $[(HSiO_{3/2})_n]$, can be converted through oxidation and/or hydrolysis to silica at relatively low temperatures. Prepared under certain conditions, the silica can be highly porous and granular in nature. For this reason a barrier layer of silicon carbide from vapor deposition of silacyclobutane is required to prevent water and chloride ion migration through the pores. The primary purpose of this layered system is to prevent corrosion of aluminum bond pads and metallization on the chip in microelectronic devices.

In the study reported here, it was of interest to test the feasibility of using the hydrogen silsesquioxane derived silica as a matrix or skeletal framework to hold silicone in place. It is postulated that the higher modulus silica in direct contact with the substrate could be used to obtain a "tighter" interface to prevent lifting of the silicone from the substrate that results in blistering. In addition, the plugging and hydrophobing of the pores in the silica with the silicone would block the transport of water and ions necessary for corrosion. It was thought that this silica/silicone nanocomposite (heterophasic at nanometer dimensions) would take advantage of the favorable properties of the individual components while eliminating the deficits of each.

In order to test the above nanocomposite system, electronic devices were coated and then exposed to salt fog. The devices were first coated with hydrogen silsesquioxane which was subsequently converted to silica. The silica coating was then vacuum impregnated with a series of polydimethylsiloxanes that varied in viscosity. To further demonstrate this corrosion protection concept for more general applications, nanocomposite films on aluminum panels were prepared and the panels were subjected to salt fog testing. The present report reviews the results of these studies.

RESULTS AND DISCUSSION

Corrosion Protection of Electronic Devices

A. Salt Fog Exposure Testing

In order to evaluate the corrosion protection by silica/silicone nanocomposites, electronic devices were coated and then subjected to an accelerated environmental test of salt fog exposure. The electronic device used for this evaluation was the Motorola 14011B CMOS chip that was mounted in an uncovered ceramic package using Au/Si eutectic die-attach. The chips had aluminum leadwires attached to gold-surfaced external bond pads. The electronic devices were first coated with hydrogen silsesquioxane that was subsequently converted to a silica-like protective film. The silica coating was then vacuum impregnated with polydimethylsiloxane or silicone fluids that varied in their viscosities. The devices were exposed to salt fog according to the MIL-STD 883C test method. Device functionality after each test interval was tested using a Pragmatic Test Systems parametric tester

Table I summarizes the results of the salt fog test. The results are shown graphically in Figure 1. All of the controls (devices coated with silica alone) had failed after 3 hour exposure while all of the silicone infiltrated samples survived up to this test interval. The results at the 3 hour test interval indicate that the infiltration of the silica framework with 100 cs silicone fluid extends corrosion protection beyond that of the silica alone. However, by the the 50 hour test interval only 30% of the devices coated with silica infiltrated with 100 cs silicone fluid were still functional and only 10% of these devices survived 100 hours of exposure. Similar to the results of use of a low viscosity fluid, a drop in corrosion protection was seen after 100 hours of exposure when the silica was impregnated with higher viscosity fluids (10000 and 600000 cs). The devices in which the silica was impregnated with 1000 cs silicone fluid had no failures even after 250 hours of salt fog exposure. The test was stopped after 250 hours because the pins on the device package were becoming so severely corroded that testing the functioning of the device would no longer be possible. This maximum in corrosion protection when the silica was infiltrated with an intermediate viscosity fluid may indicate that the lower viscosity fluid was too mobile and did not remain in the pores whereas the higher viscosity fluids may not have been able to penetrate into the pores. The 1000 cs fluid may be of the right viscosity that it can penetrate into the pores but remain in the silica framework.

In order to show that the results were not due to the silicone fluid alone, 20 CMOS 4011 devices were coated with 10,000 cs silicone fluid alone and exposed to the salt fog test. All twenty devices failed within one hour of salt fog exposure.

B. Microscopic Examination

Devices that had been exposed to salt fog were examined by optical microscopy. Bond pad regions of a CMOS 4011 device where an aluminum leadwire is attached to a metallized pad were examined at magnifications up to 1000X. Devices that had been coated with hydrogen silsesquioxane that was subsequently converted to silica at 250 °C were examined after just three hours of salt fog exposure and were found to be extensively corroded. The same bond pad regions of devices that had been coated with hydrogen silsesquioxane, converted at 250 °C, and then had silicone fluid (1000 cs) impregnated into the silica coating were also examined at magnifications up to 1000X. In this case, however, the devices had been exposed to salt fog for fifty-four hours prior to examination by optical microscopy. No corrosion of the aluminum lead wires or the bond pads had occurred.

Table I. Salt Fog Exposure Results

Device Coating	Percent Surviving				
	1 hr	3 hr	50 hr	100 hr	250 hr.
Controls, n=20 (silicone fluid only)	0	0	0	0	0
Controls, n=11 (silica only)	73	0	0	0	0
Silica/100 cs, n=10	100	100	30	10	0
Silica/1000 cs, n=9	100	100	100	100	100
Silica/10000 cs, n=10	100	100	90	70	20
Silica/600000 cs, n=10	100	100	100	80	20

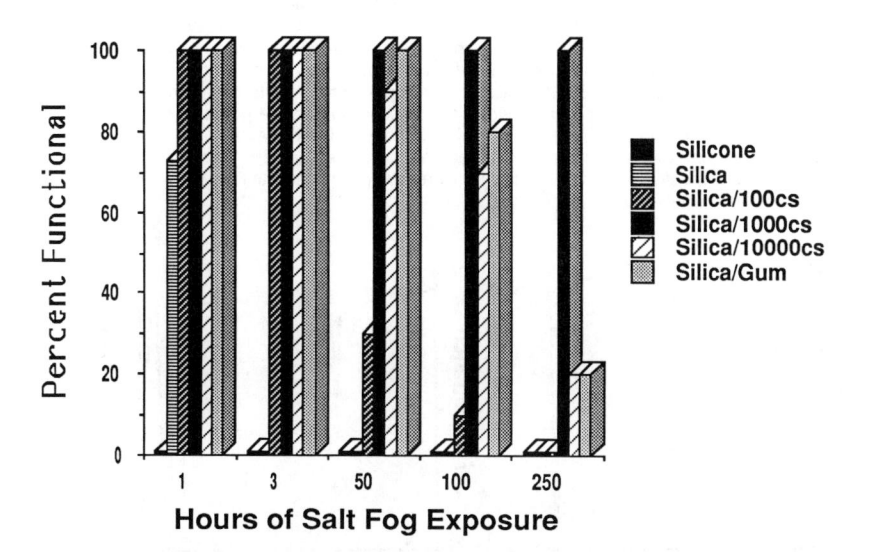

Figure 1. Salt Fog Exposure Testing of CMOS 4011 Electronic Devices. The percent of the devices still functional at the indicated test intervals are shown.

Corrosion Protection of Aluminum Panels

The results of the salt fog exposure of electronic devices suggested that it may be best to impregnate the silica with a low viscosity fluid and then subsequently crosslink or cure the fluid in place to ensure good penetration into the framework without subsequent migration of the fluid out of the structure.

In order to demonstrate the corrosion protection of aluminum by the nanocomposite system in which the silica is impregnated with low viscosity silicone fluid that is subsequently crosslinked, aluminum panels (approximately 1 inch by 3 inches) were coated. One panel remained uncoated. Another panel was coated with just silica derived from hydrogen silsesquioxane. Two panels were coated with two different formulations of silicones with SiH and Si-vinyl functionality. The silicones were crosslinked after application to the substrates by a platinum catalyzed hydrosilylation reaction. Two other panels were each coated with silica and were then impregnated with the two crosslinkable silicones. These last two panels received a heat treatment at approximately 150 °C to crosslink the silicone after infiltration. The six panels were subjected to salt fog and checked periodically by visual examination and by optical microscopy for corrosion. After 1000 hours of salt fog exposure, the aluminum panels protected by the nanocomposites consisting of silica impregnated with the crosslinked silicone showed no corrosion or loss of film adhesion. All the control samples (uncoated aluminum, aluminum coated with silica, aluminum coated with the two different formulations of silicone) showed signs of corrosion before 90 hours of salt fog exposure.

Fourier Transform Infrared Analysis

In order to demonstrate that the silicone fluid was actually infiltrated into the silica film and not just present on the surface of the film, a study using FTIR analysis was carried out. Hydrogen silsesquioxane coatings were applied to silicon wafers. The coatings were converted to silica at 250 °C. FTIR spectra of the converted coatings were then obtained, an example of which is shown in Figure 2a. A vacuum infiltration of silicone fluid (100 cs) into the silica films was then conducted. Excess fluid was then removed from the surface of the films by flooding of the wafer with toluene and spinning the wafer on a spin coater while gently applying a swab across the surface. This swabbing of the surface while spinning was repeated a second time to ensure complete removal of the silicone fluid from the film surface. FTIR spectra of the films impregnated with silicone fluid were then obtained for comparison to the silica spectrum. Evidence for the presence of the silicone fluid in the pores of the film is seen in Figure 2b in which additional peaks due to the silicone fluid are seen in the spectrum. The conclusion from these FTIR studies of nanocomposite after solvent washing was that the silicone fluid was mostly found within the silica framework and not just on its surface.

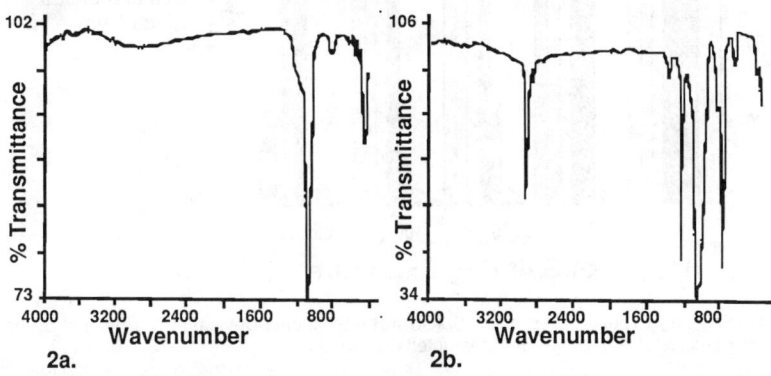

Figure 2. FTIR spectra for silica derived from hydrogen silsesquioxane (2a) and silica infiltrated with 100 cs silicone fluid (2b).

CONCLUSIONS

The present study demonstrated that nanoporous silica such as silica derived from hydrogen silsesquioxane at low temperatures can be vacuum infiltrated with solutions of silicone gum or neat silicone fluids to produce a silica/silicone nanocomposite. The examples in this study clearly showed the advantage of this nanocomposite protection system. It was shown that while neither component (silicone fluid or silica) alone provided extended salt fog protection, the two together as a nanocomposite provided greater than a 100 fold increase in salt fog protection relative to devices coated with silica. Low viscosity fluids were presumed to migrate out of the silica structure and therefore a loss in corrosion protection after a period of time was seen. On the other hand, the high viscosity fluids were thought to not penetrate into the pores of the silica and therefore offered less protection than infiltration of intermediate viscosity fluids. A maximum in the corrosion protection occurred with infiltration of intermediate viscosity (1000 cs) silicone fluid suggesting that it may be best to impregnate the silica with a low viscosity fluid and then subsequently crosslink or cure the fluid in place to ensure good penetration into the framework while preventing migration of the fluid out of the structure. Corrosion protection by a silica/silicone nanocomposite that was formed by crosslinking of the silicone after infiltration into the silica pores was demonstrated on aluminum panels.

REFERENCES

1. L.C. Wagner, Elect. Packag. Corros. Microelectron. Proc., ASM's Conf. Electron. Packag., 275-282, Ed. M.E. Nicholson, ASM Metals Park, Ohio (1987).

2. J.E. Anderson, V. Markovac and P.R. Troyk, Mater. Res. Soc. Proc., 108 (Electron. Packag. Mat. Sci. 3) 219-23 (1988).

3. J.E. Anderson, V. Markovac and P.R. Troyk, IEEE Trans. Compon. Hybrids, Manuf. Techol., 11(1), 152-8 (1988).

4. L.A. Haluska, K.W. Michael, L. Tarhay, US Patent No. 4749631.

5. L.A. Haluska, K.W. Michael, L. Tarhay, US Patent No. 4756977.

SOL-GEL SYNTHESIS OF CERAMIC-ORGANIC NANO COMPOSITES

HELMUT K. SCHMIDT
Institut für Neue Materialien, Im Stadtwald, Bldg. 43, D-6600 Saarbrücken 11, Federal Republic of Germany

ABSTRACT

Sol-gel synthesis can be used for generating nano particles from a variety of compositions. In order to avoid aggregation and undesired gelation, it is necessary to react the particle surfaces with ligands reducing their interaction but allowing the incorporation into desired matrices. Three examples (semiconducting and metal quantum dots, ZrO_2 and Al_2O_3 containing composites) are described, their properties and some applications for optics and protective coatings are discussed.

INTRODUCTION

The synthesis of inorganic-organic composites has gained increasing interest for a variety of authors [1 - 11]. Reasons for that are the possibility of developing new and interesting material properties and aspects for application [12]. Various types of structures have been proposed or identified by appropriate analytical tools. Ceramers as developed by J. Mark and G. Wilkes are prepared from elastomers by diffusion of ceramic precursors into the polymeric networks and show phase separation effects resulting in oxide or hydroxide particles of SiO_2 or TiO_2 surrounded by organic polymeric chains. Other techniques lead to interpenetrating network type structures (polyceram) [13], but structural characterization is difficult in these cases. In ORMOCER type of materials, inorganic backbones are modified by organic groupings. Polymerizable groupings can be used for crosslinking with organic monomers [14]. In some cases molecular crosslinking of the different components seem to be the most probable structure [15], especially concerning systems based on SiO_2, phenyl silanes or methyl vinyl silanes. Systems derived from other alkoxides more likely develop phase separated structures containing nano particles, as, for example, described by Naß et al. for the fabrication of monoliths based on ZrO_2 precursors [16]. If materials for optical applications are taken into consideration, phase separation has to be observed very thoroughly in order to obtain sufficient transparency and to keep scattering as low as possible.
In other cases it is of interest to generate particles of special sizes to obtain special properties, e. g. quantum effects of semiconductor dots. In these cases, especially if narrow size distribution is desired, the particle growth in the sol-gel reaction has to be controlled very carefully.
Sol-gel processing can provide the means to control particle size and phase separation and the technique allows organic groupings to be introduced for modification of the sol-gel derived materials. In this case, a variety of synthesis

routes has been developed, ranging from ≡Si-C≡ modified inorganic networks to organics linked by complex formation to inorganic units, intercalation compounds with clays or elastomers reinforced by nanoscaled sol-gel derived inorganic particles [17 - 21]. In this paper, recent developments based on ORMOCERs [22, 23], leading to novel inorganic-organic nano composites are summarized and special properties are investigated. Three examples of material development will be discussed: semiconductor quantum dot fabrication by sol-gel techniques, ZrO_2 and Al_2O_3 containing coating materials.

MATERIAL DEVELOPMENT STRATEGIES

The synthesis of glasses and ceramics by sol-gel techniques can be started from two basically different types of precursors. There are molecular precursors such as alkoxides, salts, soluble oxides or carboxylates and polymeric or colloidal precursors, which are more or less particulate. Molecular precursors very often, if no special chemistry (like the use of chelate formers blocking reactive sides) is applied, lead to colloids (SiO_2 under acid conditions is an exception) and gel formation takes place by a condensation reaction between the colloidal particles. If the structure of these gels is based on units (particles, pores, liquids in pores) in the nano range, the gels are, as a rule, of high transparency. In most cases, however, this state of homogeneity is lost by ageing, aggregations or other mechanisms leading to structural rearrangements, as schematically shown in fig. 1.

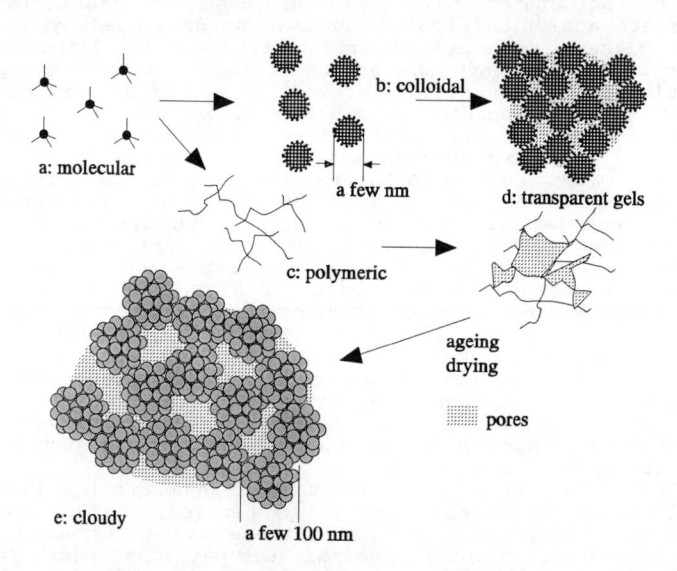

Fig. 1: Formation and coarsening of gels leading to larger structural units accompanied by loss of transparency.

As already mentioned, it is necessary to maintain the structural dimensions on a level indicated in fig. 1c, especially when optical applications are envisaged or special properties related to the nano scale shall be generated. Stabilization of this state of dispersion can be carried out easily in the liquid sol state by electrostatic stabilization by generating a charged double layer around the particles, which is well-known, for example, for ceramic slips or boehmite sols stabilized with HNO_3. But if compact materials shall be prepared from these sols, a destabilization step (e. g. pH change) has to be carried out leading to the problems indicated in fig. 1.

In order to maintain the colloidal particle properties within compact materials and to avoid structural units leading to disturbing optical properties, a new conception using colloid stabilization by well defined ligands has been developed. It is well-known that ceramic slips or suspensions can be stabilized not only by charging them electrically (ξ potential) but also by the adsorption of polymeric organic compounds to the surface. This route can be applied successfully with particles in the μm range since the volume of an adsorbed layer still remains relatively small compared to particle volume. The advantage of the adsorption of polymers is the fact that one molecule can react with more than one active site, leading to high overall adsorption coefficients. This concept cannot be used with nano scale particles if high package densities are required since the volume fraction of the particle decreases drastically (fig. 2).

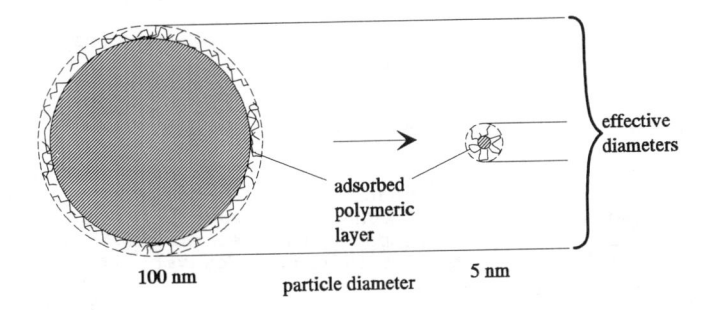

Fig. 2: Comparison of the effective diameters of polymer stabilized particles with two different particle diameters.

The use of ligands interacting specificly with the colloid surface enables the use of low molecular weight compounds thus reducing the "ligand volume". In addition to this, very specific properties can be obtained by proper choice of the ligands, too. This is of special interest if ligands with reactive properties are chosen to be used for anchoring the particles within a matrix. In our case, ORMOCER matrices have been investigated, since the ORMOCER system offers a high flexibility with respect to matrix composition and processing techniques. The strategy of the composite development is given in fig. 3. It includes the synthesis of small particles, the stabilization of the particles by ligand/particle interaction and the incorporation into ORMOCER matrices.

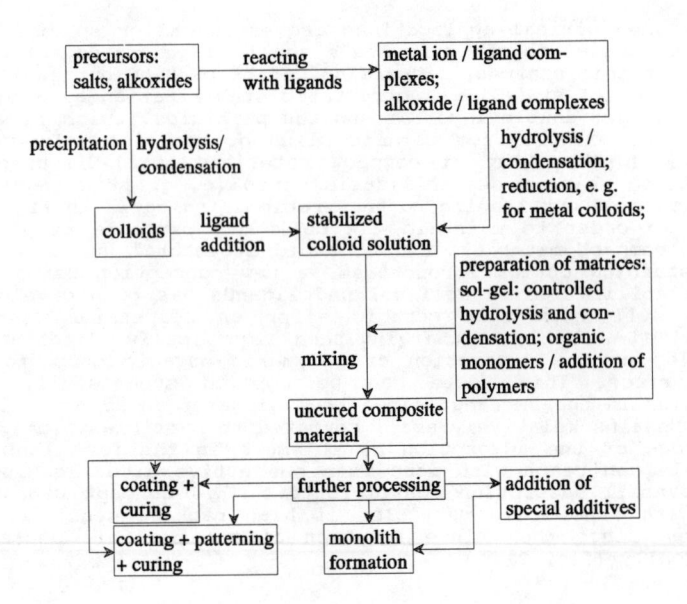

Fig. 3: Draft of sol-gel preparation of nano composites using ligand interaction stabilization.

QUANTUM DOT CONTAINING COMPOSITES

Semiconductor and metal quantum dots have become of high interest due to their potential for χ^3 applications. As shown by Henglein et al. [24], CdS and other II/VI semiconductor colloids can easily be prepared by precipitation from aqueous solution using Cd salts and sulfides. However, it is difficult to produce compact materials due to the problems mentioned above. Other work has been reported on generation of sol-gel derived CdS/glass composite by impregnation of gels by Cd salts or simultaneous precipitation [25, 26] or by phase separation from melted glasses. But it is difficult to obtain narrow size distributions of the colloids by these methods. The application of methods indicated in fig. 3 led to stable sols with narrow size distributions (indicated by the optical spectra) for a variety of systems. Details of the synthesis procedures have already been described by Spanhel and coworkers for CdS. For quantum dot systems and the incorporation of the dots into ORMOCER matrices the following processes have been developed [27] (fig. 4 and fig. 5):

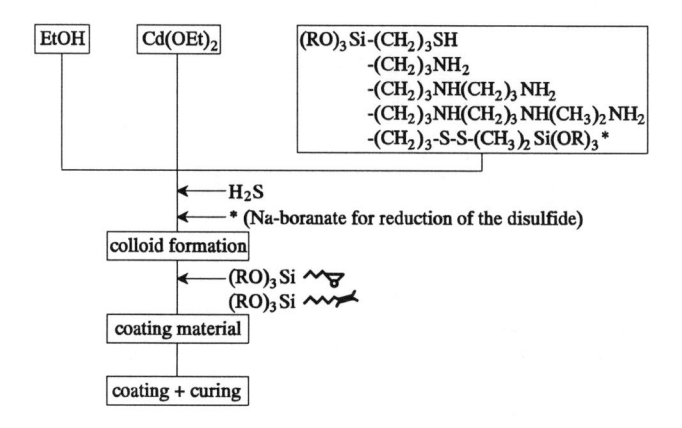

Fig. 4: Flow diagram of the preparation of ligand stabilized CdS quantum dots.

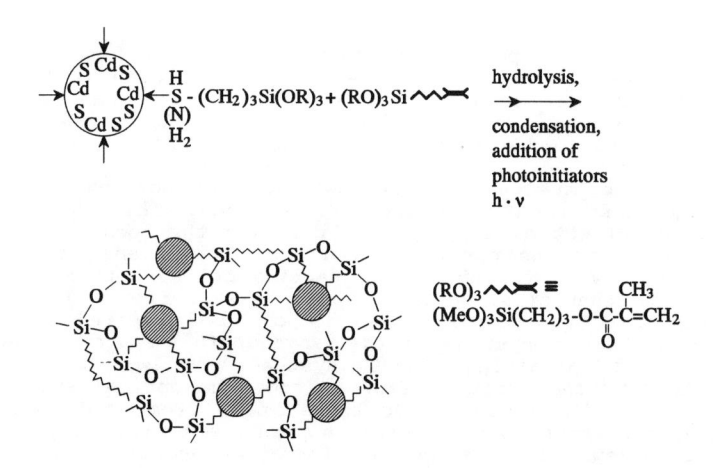

Fig. 5: Reaction and structure model of the incorporation of ligand stabilized CdS quantum dots into ORMOCER matrices.

In fig. 6 UV/VIS spectra of CdS containing materials are shown.

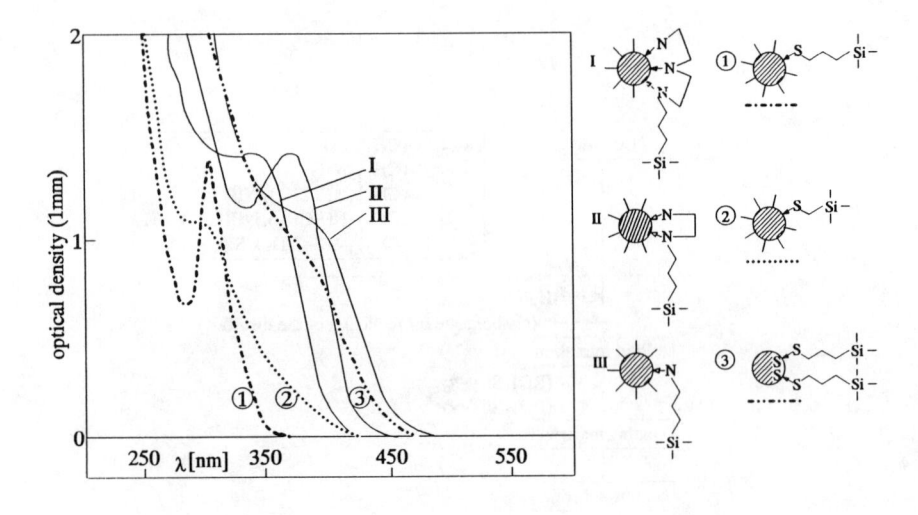

Fig. 6: Absorption spectra of sulfide and ammine stabilized CdS quantum dots.

The optical spectra show that there is a relatively sharp cut off due to the band gap energy of the quantum dots. At smaller particle diameters, sharp exitonic peaks can be detected in the sulfide stabilized systems. TEM investigations confirm particle diameters ≤ 2 nm, but no quantitative analysis is possible.

The spectra of the colloids do not change after incorporating them into ORMOCER matrices. This proves the shielding effect of the ligands, defining the local electronic environment at the particles' surface. This means that it is possible to vary the optical properties of the dots by varying the ligand chemistry, which can also be seen from fig. 6. In a similar way, PbS coated CdS dots can be prepared [28]. χ^3 values of this system are in the range of 10^{-9} esu at present.

Gold containing ORMOCER systems have been synthesized according to the draft in fig. 4, too. The optical spectra of these systems are similar to those known from gold colloids. The incorporation into ORMOCERs is possible by the bifunctional ligand concept [29]. The colloid formation process in films was optimized in a way that gold complexes were introduced into a viscous, partially condensed ORMOCER system containing polymerizable ligands. After the preparation of films on glass substrates, the colloid formation process can be initiated by UV light. The colloid growth process takes place at T ≈ 100 °C during the irradiation used for photopolymerization curing. The fabrication is schematically shown in fig. 7.

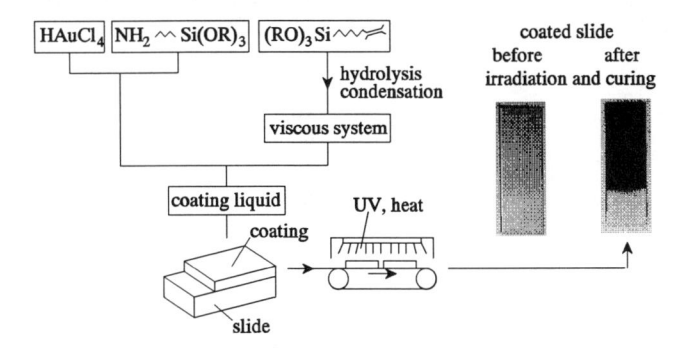

Fig. 7: Draft of the one step preparation of gold colloids in ORMOCERs. T ≈ 100 °C, $t_{irr.}$ = 15 min.

Experimental details about the complex and ORMOCER synthesis are given in [29]. The films can be prepared with high gold contents up to several volume percents and optical densities up to 5. In the four wave mixing experiment the films show a self-diffraction effect attributed to χ^3 NLO properties, but no quantitative evaluation has been carried out so far. The results demonstrate the potential of the sol-gel technique for the fabrication of optical nano composites with interesting properties.

SYNTHESIS AND UTILIZATION OF ZrO_2 PARTICLE CONTAINING COMPOSITES

ZrO_2 in ORMOCERs has been investigated by Dislich [30] for increasing the refractive index of these materials. Naß has shown that high ZrO_2 contents can be obtained by using ß-diketones as complex formers for Zr alkoxides and ORMOCERs as matrix [16]. He found particle sizes of about 10 nm. Sanchez [31] showed that the cluster size of Zr-alkoxides complexed by ß-diketones can be varied between 5 and 50 nm, and Rinn showed that this can be done up to 3 µm [32]. In our investigation we focused on the complexation with methacrylic acid (MAS) due to their ability for undergoing a polymerization reaction. As already shown in [33], complexation of MAS leads to the formation of nano particle containing sols the size of which depends on the ratio r = Zr:MAS. While between r = 1:1.6 and 1:1.2 the particle diameter d remains at about 3 nm, lower r lead to a strong increase of d and above r = 1:0.8, precipitation takes place. But as shown by Wilhelm and coworkers [34], nano scaled clusters can be obtained with r ≈ ≤ 4, if "water" is introduced into the system by ≡SiOH groups, e. g. using prehydrolysed $(RO)_3Si(CH_2)_3OCOC(CH_3)=CH_2$. The particle size in this case mainly depends on the concentration of the "latent" water introduced into the system through the silanes. In fig. 8, the particle size development is shown as a function of composition and time. For the synthesis the process described in [35] was used. As one can see from fig. 8, at first a bimodal distribution occurs, equilibrizing into a monomodal distribution after several days. The reasons for that probably are structural rearrangements the mechanisms of which are not quite clear so far.

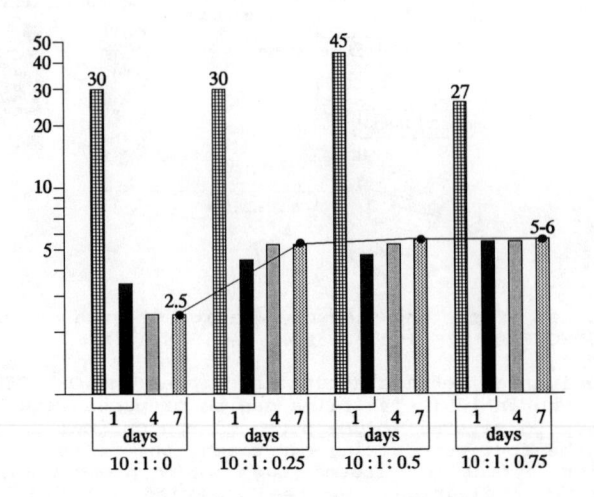

Fig. 8: Particle size distribution in liquid uncured composite materials after [29]. a:b:c = methacryloxy silane: Zr propylate:MAS; days: reaction time at room temperature after mixing the (complexed) alkoxide with the silane prereacted after [30].

The tendency of increased particle sizes by increasing the ratio $x = (b+c)/a$ was supported by an experiment using the ratio a:b:c = 10:3:2.25, ending up with a particle size of about 15 nm. The particle analysis was carried out by photon correlation spectroscopy. The analysis of the data shows almost monodispersed systems and spherical particles.

Mechanical tests carried out on cured coatings show a good scratch resistance of these systems (1 - 2 % haze after the standard test with a taber abrader). The refractive index can be varied between 1.52 and 1.58. The most interesting properties of the system, however, are their photoresist behavior and their ability to adjust the viscosity as a function of the processing condition. It can range from some mPa·s to almost solid systems in the uncured form, being able to be embossed during the final curing step. The resist behavior was used for patterning of planar waveguides, as described in [33 and 35]. Embossing can be carried out using a UV source and a transparent substrate. The photopolymerization occurs during the embossing step. Photolithographic method (laser writing, maskaligner techniques, beam energy fluctuation by two wave mixing interferences) have been developed for patterning, too. The irradiated areas get polymerized and the non-irradiated areas can be redissolved by diluted NaOH or organic solvents. The two wave mixing experiment is of interest due to the possibility of continuing with one wave and thus "writing" incoupler gratings and channel waveguides in one step. In fig. 9 the schematics of the two wave mixing writing and a grating produced by this technique is shown [36].

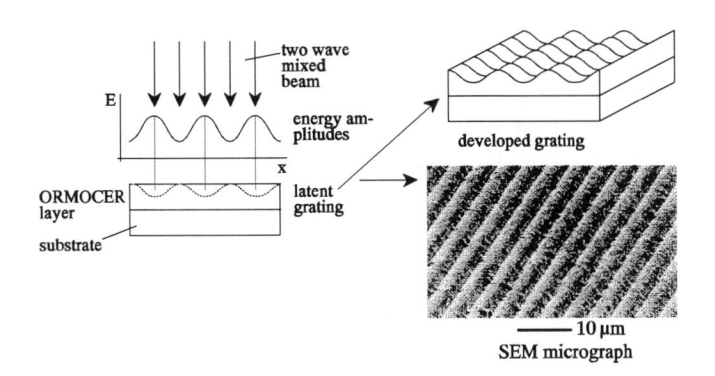

Fig. 9: Fabrication of a diffraction grating by a two wave mixing photolithography process [after 36].

Optical losses of the system are in the range of ≈ 1 dB/cm without using clean room conditions at present. The nano composite seems to be a suitable matrix for processing of optical waveguides. They can be doped either with push-pull molecules (dyes), lanthanides or quantum dots, which already has been proved experimentally.

Al_2O_3 CONTAINING COMPOSITES FOR COATINGS

As described elsewhere, transparent hard coatings have been developed for plastics [12] and brass protection [37], but no structural analysis was carried out. Since Al containing composites have shown a good corrosion protection for brass, the question arose how far corrosion protection for other metal surfaces can be obtained and how far structures can influence the properties. In [38] a system synthesized from $(RO)_3Si(CH_2)_3OCH_2-\overline{CH-CH_2O}$, $(RO)_3Si(CH_2)_2CH_3$ and $Al(oBut)_3$ is described and experimental details are given. The epoxide is polymerized with methyl imidazole as catalyst. In order to avoid precipitation, the Al butylate was reacted with ethoxybutanol to form a complex according to [39]. Details of the reaction are given in [36]. The complexed system was reacted with the silanes according to fig. 8. Different types of catalysts were used. Only F^- catalysis leads to the formation of non-spherical particles which cannot easily be explained. Further investigations by photon correlation spectroscopy show that particles can be obtained from the silane precursors without Al, too, but they are nearly spherical. In fig. 10 the particle size development in the three component system is shown as a function of time.

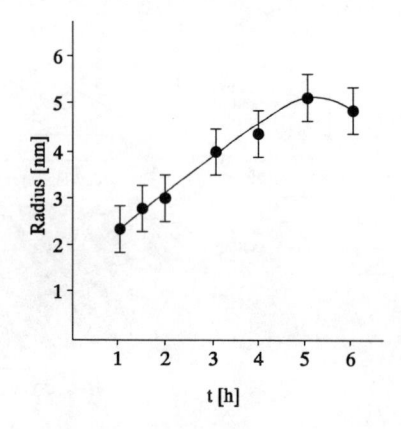

Fig. 10: Development of nano particles in the three component system according to fig. 8 after [36], method: photon correlation spectroscopy.

So far, the mechanisms for these complex reactions are not clear, but the effect on the nano particle containing systems as coatings on Al sheets is surprising: Only the particle containing system shows the highest corrosion protection effect on the Al surface combined with a very good scratch resistance. No traces of corrosion can be detected after two weeks of salt spray test at 35 °C and 100 % r.h. While the high scratch resistance can be attributed to the nano particles, for the increased corrosion resistance no simple explanation exists so far. But the results show that by development of nano scaled particles within ORMOCERs interesting properties can be generated.

CONCLUSION

The investigations show that by specific chemical means nano sized particle containing composites can be generated. The chemistry is mainly based on sol-gel or related techniques. It seems to be of high importance to tailor the interface between the particles and the matrix. This influences the optical properties and probably the particle distribution, too. But in any case it is necessary to avoid undesired reactions between the particles leading to aggregation.

ACKNOWLEDGEMENT

The author wants to thank Dr. Gerhard, Dipl.-Chem. Wagner, Dr. Spanhel, Dr. Merl and Dr. Krug for their helpful discussions and the Minister for Research and Culture of the State of Saarland for his financial support.

REFERENCES

[1] G. L. Wilkes, B. Orler and H. H. Huang, Polym. Prepr 26, 300 (1985).
[2] L. Garrido, J. L. Ackermann, J. E. Mark, Mat. Res. Soc. Symp. Proc. 171, 65 (1990).
[3] H. Schmidt, in Chemical Processing of Advanced Materials; edited by L. L. Hench and J. K. West (J. Wiley & Sons, New York, 1992).
[4] B. M. Novak and C. Davies, Polym. Prepr. 32, 512 (1991).
[5] C. J. T. Landry and B. K. Coltrain, Polym. Prepr. 32, 514 (1991).
[6] H. H. Huang, B. Orler and G. L. Wilkes, Polym. Bull. 14, 557 (1985).
[7] H. H. Huang, B. Orler and G. L. Wilkes, Macromolecules 20, 1322 (1987).
[8] A. B. Brennan and G. L. Wilkes, Polymer 32, 733 (1991).
[9] C. Sanchez, in Proc. 6th International Workshop on Glasses and Ceramics from Gels, October 1991, Sevilla/Spain, edited by L. Esquivias (J. Non-Cryst. Solids), in print.
[10] J. D. Mackenzie, in: Proc. 6th International Workshop on Glasses and Ceramics from Gels, October 1991, Sevilla/ Spain edited by L. Esquivias (J. Non-Cryst. Solids), in print.
[11] P. N. Prasad, in: SPIE Vol. 1328 Sol-Gel Optics, edited by J. D. Mackenzie and D. R. Ulrich (SPIE, Bellingham/ Washington, 1990), p. 168.
[12] H. Schmidt, B. Seiferling, G. Philipp and K. Deichmann, in: Ultrastruacture of Processing of Advanced Ceramics, edited by J. D. Mackenzie and D. R. Ulrich (John Wiley & Sons, New York, 1988), p. 651.
[13] G. Teowee, J. M. Boulton, H. H. Fox, A. Koussa, T. Gudgel and D. R. Uhlmann, Mat. Res. Soc. Symp. Proc. 180 (1980).
[14] H. Schmidt, Mat. Res. Soc. Symp. Proc. 171, 3 (1990).
[15] H. Schmidt, H. Scholze and G. Tünker, J. Non-Cryst. Solids 80, 557 (1986).
[16] R. Naß and H. Schmidt, in: Ceramic Powder Processing Science, edited by H. Hausner, G. L. Messing and S. Hirano (Deutsche Keramische Gesellschaft, Köln), p. 69.
[17] H. Schmidt, Mater. Res. Soc. Symp. Proc. 32, 327 (1984).
[18] H. Krug and H. Schmidt, in: Proc. Workshop "Integrated Optics and Microoptics with polymers", Mainz/FRG, March 1992 (in print).
[19] L. C. Klein and B. Abramoff, Polym. Prepr. 32, 519 (1991).
[20] B. S. Dunn, J. D. Mackenzie, J. I. Zink, O. M. Stafsudd, in: SPIE Vol. 1328 Sol-Gel Optics, edited by J. D. Mackenzie and D. R. Ulrich, 1990, p. 174.
[21] A. Okada, K. Fukumori, A. Usuki, Y. Kojima, N. Sato, T. Kurauchi and O. Kamigaito, Polym. Prepr. 32, 540 (1991).

[22] H. Schmidt, in: **Proceedings of the International Symposium on Molecular Level Designing of Ceramics**; edited by the Team of the NEDO International Joint Research Program, Nagoya, 1991, p. 59.

[23] H. Schmidt, H. Krug and N. Merl, Rivista della Staz. Sper. Vetro 23, 11 (1992).

[24] A. Henglein, Topics in Current Chemistry: Mechanism of reactions on colloidal microelectrodes and size quantization effects 143, 115 (1988).

[25] C. Li, Y. J. Chung, J. D. Mackenzie and E. T. Knobbe, presented at Amer. Cer. Soc. Optical Materials Symposium, October 1991, Washington/USA.

[26] K. Fukomi, A. Chayahara, K. Kadono, T. Sakaguchi, Y. Horino, J. Hayakawa and M. Satou, Jpn. J. Appl. Phys. 30, L 742 (1991).

[27] L. Spanhel, E. Arpaç and H. Schmidt, in: Proc. 6th International Workshop on Glasses and Ceramics from Gels, edited by L. Esquivias (J. Non-Cryst. Solids) in print.

[28] L. Spanhel, H. Schmidt, A. Uhrig and C. Klingshirn, in: Proc. MRS Spring Meeting 1992, San Francisco (Mat. Res. Soc. Symp. Proc.) in print.

[29] L. Spanhel, M. Mennig and H. Schmidt, in: Proc. ICG Congress on Glass, Madrid, Oktober 1992, accepted for publication.

[30] B. Lintner, N. Arfsten and H. Dislich, J. Non-Cryst. Solids 100, 378 (1988).

[31] C. Sanchez, in: Proc. 2nd Eurogel Conf., June 1991, Saarbrücken, North Holland Publishers (in print).

[32] G. Rinn und H. Schmidt, in: **Ceramic Powder Processing Science**, edited by H. Hausner, G. L. Messing and S. Hirano (Deutsche Keramische Gesellschaft, Köln), p. 221

[33] H. Schmidt, H. Krug and R. Kasemann, in: **Homage** to Galileo, edited by P. Mazzoldi (Università di Padova, 1992) p. 303.

[34] R. Wilhelm and coworkers, private communication.

[35] H. Schmidt, H. Krug, R. Kasemann, F. Tiefensee, in: **SPIE Vol. 1590 Submolecular Glass Chemistry and Physics**, edited by N. Kreidl (SPIE Bellingham/Washington, 1991) p. 36.

[36] P. Oliveira and H. Krug, private communication.

[37] Annual report of Fraunhofer-Institut für Silicatforschung, 1990, p. 55.

[38] G. Wagner, Master's Thesis, Saarbrücken 1992.

[39] M. Fedtke, Makromol. Chem., Macromol Symp. 7 153 (1987).

VARIABLE FREQUENCY CONDUCTIVITY OF LAYERED POLYPYRROLE / V$_2$O$_5$ COMPOSITES

D. C. DeGROOT [a], J. L. SCHINDLER [a], C. R. KANNEWURF [a],
Y.-J. LIU [b], C.-G. WU [b], and M. G. KANATZIDIS [b]

(a) Department of Electrical Engineering and Computer Science and the Materials Research Center, Northwestern University, Evanston, IL 60208
(b) Department of Chemistry and the Center for Fundamental Materials Research, Michigan State University, East Lansing, MI 48824

ABSTRACT

The frequency dependent electrical properties of the intercalated polypyrrole/V$_2$O$_5$ system have been measured. This study continues the investigation into the charge transport mechanisms that have been identified in these layered polymer/inorganic composites. The polypyrrole/V$_2$O$_5$ material is prepared by *in-situ* oxidative polymerization of pyrrole in the intralamellar space of the V$_2$O$_5$ xerogel. This process produces a layered, two-dimensional structure in which the charge transport properties result from two parallel conduction paths: the polypyrrole chains and the V$_2$O$_5$ layers. Impedance spectroscopy data have been collected from free-standing film samples of the layered polypyrrole/V$_2$O$_5$. The experiments were conducted over the frequency range of 10 Hz to 0.5 GHz and a temperature range of 77 to 310 K. The polypyrrole/V$_2$O$_5$ impedance results are presented in various equivalent forms and compared to data collected from pristine V$_2$O$_5$•nH$_2$O, and reduced Cs$_{0.14}$V$_2$O$_5$•nH$_2$O.

INTRODUCTION

This research program has focused on stable, electrically conductive polymers such as polypyrrole (PPY), polythiophene (PTP), and polyaniline (PANI) since these polymers exhibit high electrical conductivities (σ). These materials also show considerable promise for use in device applications and electrostatic shielding.[1] The study of anisotropic charge transport in this class of materials is simplified if the polymer chains are ordered in one or more dimensions. In addition, it has been shown that electrical conductivity is enhanced significantly with chain ordering.[2] Most electrochemical and chemical preparation processes, however, produce amorphous forms of the conducting polymers, and it is primarily the electrochemical methods that have proven effective in fabricating thin film forms suitable for applications.

Investigation into the intercalation properties of V$_2$O$_5$•nH$_2$O as a host for redox polymerization has led to the creation of various ordered, conducting polymer/inorganic composites.[3-7] These studies have shown that monolayers of long polymer chains can be inserted into the intralamellar space of V$_2$O$_5$•nH$_2$O xerogels by *in-situ* intercalation/oxidative polymerization. This chemical synthesis route yields layered films in which the conducting polymer is ordered at least perpendicular to the stacking direction. The composite materials not only provide an ordered environment for studying charge transport behavior of the conducting polymers, but may also possess unique properties not accessible in either constituent separately.

Initial reports from this laboratory on the PPY/V$_2$O$_5$•nH$_2$O system presented variable-temperature electrical conductivity and thermoelectric power data for several samples of differing PPY content.[3-5] The conductivity of these films was compared to the conductivity of pristine PPY and Na$_{0.40}$V$_2$O$_5$•1.0H$_2$O. For low polymer content, the transport properties reflected the behavior of reduced V$_2$O$_5$•nH$_2$O, and for high polymer content the conductivity and thermopower tended towards the behavior of PPY. Since both the polymer and V$_2$O$_5$•nH$_2$O layers are electrically conductive, it becomes important to identify the dominant transport mechanisms in these composites.

The charge transport studies of the conducting polymer/inorganic composites can be enhanced with variable frequency ac measurements. In particular, impedance spectroscopy can be applied to multi-component and multi-phase solids to identify the contributions of each constituent.[8] Impedance (**Z**) is a complex quantity defined as the ratio of the electric field and magnetic field vectors in a medium. The impedance of a material represents the magnitude and phase of the voltage across the sample divided by the magnitude and phase of the current flowing through the sample. Impedance data are often collected by sweeping the frequency of a sinusoidal stimulus over a specific range of interest. When these data are plotted in the impedance plane as the imaginary part (X) vs. the real part (R), the locus of points frequently forms semi-circular arcs. Multiple arcs are found in many electronic multi-phase materials, each arc representing a different phase or boundary region. The electrical parameters of each phase can be identified by fitting the data to appropriate equivalent circuit models. For materials with a frequency dependent conductivity and permittivity, the impedance spectra form distorted semicircular arcs. In such cases it is possible to calculate the complex conductivity and permittivity from the impedance data and plot these quantities as functions of frequency as well.

Impedance spectroscopy has been applied to the PPY/V_2O_5•nH_2O system to further investigate charge transport in these composites. This report presents the initial variable temperature and variable frequency electrical properties for two composite (PPY)$_x$$V_2O_5$•$nH_2O$ films: x=0.73 and x=1.17. The frequency dependent conductivity of the composite structures are compared to a chemically reduced V_2O_5•nH_2O film—$Cs_{0.14}V_2O_5$•nH_2O.

EXPERIMENTAL

V_2O_5•nH_2O xerogels are layered porous solids and are obtained by the polycondensation of vanadic acid.[9] Solutions of freshly prepared vanadic acid undergo a sol-gel transition. As the water spontaneously evaporates it leaves behind a red hydrated oxide (V_2O_5•nH_2O) as an almost dry solid, called a xerogel. During the polymerization process some reduction unavoidably occurs and a mixed valence compound is actually obtained in which some portion of vanadium atoms are reduced to V^{4+}. The extent of reduction varies from 0.5% to 10% and is responsible for the finite small polaron n-type conductivity observed in these systems.[10-12] The PPY/V_2O_5•nH_2O compounds were prepared by the reaction of pyrrole with V_2O_5•nH_2O in water[3,4] and $Cs_{0.14}V_2O_5$•nH_2O was prepared by the action of CsI on V_2O_5•nH_2O in acetonitrile.[13]

Small sections of the self-supporting film samples were cut to be compatible with the dimensions of the coaxial sample mounting arrangement used in these electrical impedance studies (Figure 1). The thin sections (40-70 μm) were disc-shaped with an outer diameter of approximately 3.6 mm and a concentric hole of approximately 1.2 mm in diameter. The discs were electrically connected to the sample mount conductors by painting thin rings of gold colloidal paste around the inner and outer edges of the disc at the film-conductor interfaces. The contact resistance typically ranged between 1-10 Ω and accounted for no more than 2% of the total dc sample resistance.

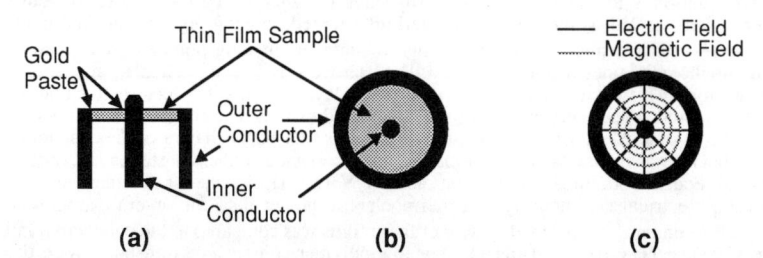

Figure 1. Coaxial sample mounting arrangement used in the variable frequency impedance measurements. (a) Cut-Away View. (b) Top View. (c) Electric and magnetic field lines in the coaxial sample.

Impedance data were collected using a computer automated network analyzer system over the frequency range of 10 Hz to 0.5 GHz and over the temperature range of 80 K to 310 K. In the temperature dependent studies, the samples were placed at the end of a cryogenic probe on a large copper sample stage. An Au(0.07% Fe)/Chromel thermocouple mounted on the sample stage directly below the samples provided accurate temperature readings. The jacket surrounding the samples was filled with dry helium gas to assure good thermal contact, and the samples were cooled to liquid nitrogen temperatures from room temperature at approximately 2°/minute. Wide-band frequency sweep measurements were performed at specified temperatures as the system warmed over a 20-24 hour period.

The impedance magnitude (|Z|) and phase (Θ) values were collected under software control using a vector network analyzer in a one-port configuration. The coaxial sample mount was attached to the end of a semi-rigid coax cable running down the cryogenic probe. By comparing the magnitude and phase of the incident and reflected waves in the transmission line, the network analyzer computed the sample impedance terminating the line. Since the electric fields are radially directed in the coaxial line and sample mount (Figure 1c), the measured response is due to transport in the PPY and $V_2O_5 \cdot nH_2O$ layers that lie in the plane of the film. Before all impedance spectroscopy experiments, the system was fully calibrated at room temperature at the same plane where the samples were to be located. Although the calibration was uncompensated for temperature dependent geometry changes, previous calibration runs have shown less than 10% error in |Z| over the temperature range of interest due to thermal expansion/contraction of the cabling and sample mounts.

Variable temperature dc conductivity measurements were made on the film samples in a direction parallel to the layers using a separate computer automated system.[14] Rectangular sections of films (typically 5x2 mm^2) were measured using a standard four-probe technique. The dc measurements provided a good base-line reference and comparison check for the frequency dependent conductivity values calculated from the impedance data.

RESULTS AND DISCUSSION

Impedance spectra were collected for $(PPY)_{1.17}V_2O_5 \cdot nH_2O$, $(PPY)_{0.73}V_2O_5 \cdot nH_2O$, and $Cs_{0.14}V_2O_5 \cdot nH_2O$ films at various temperatures. The data were first plotted in the impedance plane to check for possible multiple arcs. There was no evidence for multiple arcs in any of the data sets when impedance was measured parallel to the layers over the frequency range of 10 Hz to 0.5 GHz. The low frequency impedance values were entirely real and converged to the measured dc resistance values. A single arc formed as frequency was increased, with the values becoming entirely real and close to zero impedance at very high frequencies. The formation of a single impedance arc without low frequency tails or spurs indicates that the charge transport in these materials is primarily electronic in nature. The single arcs were not perfect semi-circles and the data could not be fit to an equivalent parallel resistor-capacitor circuit. The ideal parallel resistor-capacitor circuit model is used to describe electronic systems in which the conductivity is frequency independent with a single relaxation time. This circuit is not adequate for modeling the $PPY/V_2O_5 \cdot nH_2O$ and $Cs/V_2O_5 \cdot nH_2O$ behavior which exhibit a frequency dependent conductivity.

Complex conductivity ($\sigma^* = \sigma' + j\sigma''$) values were calculated from the impedance data by first taking the reciprocal of $Z(\omega)$ to obtain the Admittance (**Y**). The $Y(\omega)$ values were then converted to $\sigma^*(\omega)$ using the geometry factors obtained from the samples with a calibrated binocular microscope scale. The real part of the room-temperature complex conductivity is plotted against the logarithm of frequency ($\log_{10} f$) in Figure 2 for the three films investigated.

$(PPY)_{1.17}V_2O_5 \cdot nH_2O$ displays the highest conductivity of the three films. These data exhibit a frequency independent region with a value that closely matches the dc conductivity value (see Figure 4). Above 1 MHz, $\log \sigma'(\omega)$ increases nearly linearly with log f. The $\sigma'(\omega)$ data for $(PPY)_{0.73}V_2O_5 \cdot nH_2O$ seem to follow the behavior of $Cs_{0.14}V_2O_5 \cdot nH_2O$ up to approximately 5 MHz. From there the slope changes and the conductivity increases to values approaching that of $(PPY)_{1.17}V_2O_5 \cdot nH_2O$. For $Cs_{0.14}V_2O_5 \cdot nH_2O$, $\log \sigma'(\omega)$ increases linearly with log f over the frequency range of Figure 2. Frequency dependent conductivity for $(PPY)_{1.17}V_2O_5 \cdot nH_2O$ is shown in Figure 3 for four different temperatures. The frequency independent region moves to lower frequencies with decreasing temperature and the slope of the ac portion of $\log \sigma'(\omega)$ vs. log f increases slightly as temperature decreases. The scatter in the data increased for lower temperatures as the impedance of the thin disc sample approached the instrumentation limits.

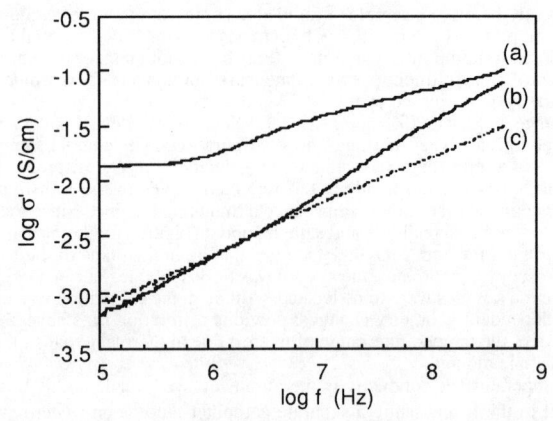

Figure 2 Real part of the complex conductivity vs. frequency at 300 K for (a) $(PPY)_{1.17}V_2O_5 \cdot nH_2O$, (b) $(PPY)_{0.73}V_2O_5 \cdot nH_2O$, and (c) $Cs_{0.14}V_2O_5 \cdot nH_2O$ films.

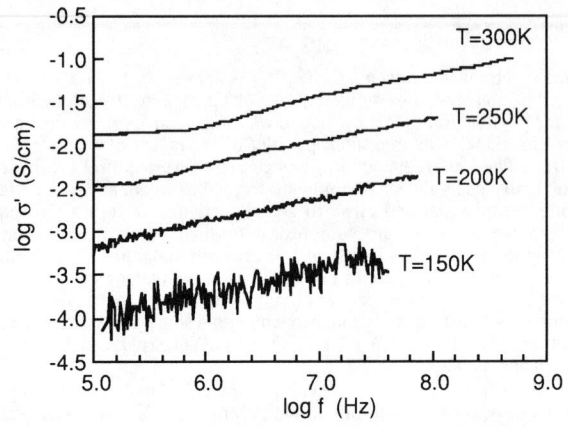

Figure 3 Real part of complex conductivity vs. frequency for the $(PPY)_{1.17}V_2O_5 \cdot nH_2O$ film at various temperatures.

Variable temperature dc conductivity data were also collected for the the three films discussed above, along with a pristine $V_2O_5 \cdot nH_2O$ xerogel film. The four-probe dc conductivity data provided a reference for the ac study. An estimate of the dc contact resistance was made by comparing the dc conductivity values with the low frequency limit of the ac conductivity. The two values were always found to be in close agreement, implying that contact resistance accounts for a small percentage of the total $\sigma'(\omega)$ quantity. The dc data are plotted in Figure 4 against reciprocal temperature. The $Cs_{0.14}V_2O_5 \cdot nH_2O$ data, omitted from Figure 4 for clarity, fall on a line between curves (b) and (c). Both the pristine $V_2O_5 \cdot nH_2O$ film and the $Cs_{0.14}V_2O_5 \cdot nH_2O$ seem to follow an activated behavior. The $(PPY)_{1.17}V_2O_5 \cdot nH_2O$ sample, on the other hand can be described better as following a $T^{-1/4}$ to $T^{-1/2}$ behavior, indicative of a hopping mechanism. As with the case of low frequency $\sigma'(\omega)$, the $(PPY)_{0.73}V_2O_5 \cdot nH_2O$ dc results are more like the $Cs_{0.14}V_2O_5 \cdot nH_2O$ conductivity than that of $(PPY)_{1.17}V_2O_5 \cdot nH_2O$.

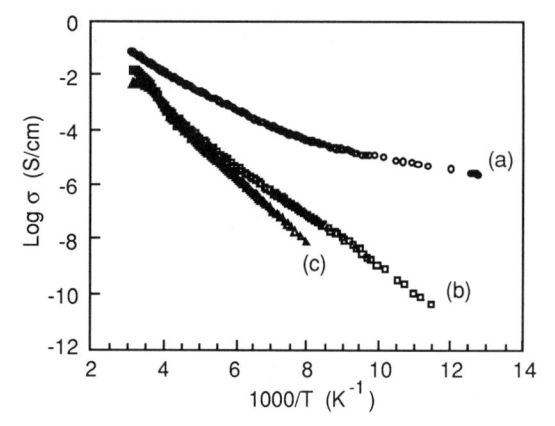

Figure 4 Temperature dependent dc conductivity for films of: (a) $(PPY)_{1.17}V_2O_5 \cdot nH_2O$, (b) $(PPY)_{0.73}V_2O_5 \cdot nH_2O$, and (c) $V_2O_5 \cdot nH_2O$.

The $\sigma'(\omega)$ data of Figures 2 and 3 are comprised of dc and ac components.

$$\sigma'(\omega,T) = \sigma_{dc}(T) + \sigma_{ac}(\omega,T) \quad (1)$$

The behavior of the frequency dependent portion is accessible by subtracting the dc or frequency independent values from $\sigma'(\omega,T)$. At constant temperature, the $\sigma_{ac}(\omega)$ behavior was found to be consistent with the the power law expression used to describe conductivity in a variety of organic and inorganic materials.[15,16]

$$\sigma_{ac}(\omega) = \sigma_0 \omega^s \quad (2)$$

where σ_0 and s are temperature dependent parameters. In the $(PPY)_{1.17}V_2O_5 \cdot nH_2O$ sample, the exponent s decreases slightly with temperature from s=0.49 at 150 K to s=0.32 at 300 K. The s value for $Cs_{0.14}V_2O_5 \cdot nH_2O$ at 300 K (s=0.54) is significantly greater than that for $(PPY)_{1.17}V_2O_5 \cdot nH_2O$ at the same temperature.

If the power law behavior of $\sigma_{ac}(\omega)$ is due to barrier hopping conductivity, s would be directly related to the hopping barrier height and inversely related to the temperature.[17] The frequency dependent conductivity for the three films and the dc data for $(PPY)_{1.17}V_2O_5 \cdot nH_2O$ seem to be consistent with such a barrier hopping model. Additional data are required to distinguish between single polaron and possible bipolaron hopping in these materials.

CONCLUSION

This initial impedance spectroscopy study of composite $PPY/V_2O_5 \cdot nH_2O$ films shows that the frequency dependent electrical behavior is not obviously separable into polymer and $V_2O_5 \cdot nH_2O$ contributions over the frequency range of 10 Hz to 0.5 GHz. Rather, the lack of multiple arcs in the impedance plane tends to support the idea that the $(PPY)_x V_2O_5 \cdot nH_2O$ materials are not a multi-phase systems. It may still be possible, however, that the two materials have very similar relaxation mechanisms, in which case the two contributions could not be distinguished. Furthermore, the formation of a single impedance plane arc with entirely real values in the low frequency limit indicates that the electrical conductivity arises from electronic processes as opposed to ionic ones. The frequency dependent data are consistent with the dc conductivity collected, and follow a power law behavior that has been used to describe hopping conductivity in a wide variety of materials. The exponent in the power law expression tends to decrease as temperature is increased for the high polymer content film, supporting a barrier hopping model. The ac and dc behavior of the low polymer content material tends to follow the behavior of the reduced $Cs_{0.14}V_2O_5 \cdot nH_2O$ film. As the polymer content is increased, the conductivity increases and the power law exponent decreases, indicating better polymer connectivity and lower barrier height with polymer content. Further investigations are now in progress to characterize the ac conductivity data perpendicular as well as parallel to the layer planes.

ACKNOWLEDGEMENTS

This work supported by the National Science Foundation: at Northwestern University through the Materials Research Center, Grant No. DMR-8821571; at Michigan State University, Grant No. DMR-8917805.

REFERENCES

1. Proceedings of the International Conference on Science and Technology of Synthetic Metals (ISCM'90), edited by M. Hanack, S. Roth, and H. Schier, Synth. Met. 41 (1991).

2. E. M. Scherr, et. al., Synth. Met. 41, 735 (1991).

3. C.-G. Wu, M. G. Kanatzidis, H. O. Marcy, D. C. DeGroot, and C. R. Kannewurf, in Lower Dimensional Systems and Molecular Devices, edited by R. M. Metzger (Plenum Press, New York, 1991) pp. 427-434.

4. C.-G. Wu, M. G. Kanatzidis, H. O. Marcy, D. C. DeGroot, and C. R. Kannewurf, Polym. Mat. Sci. Eng. 61, 969 (1991).

5. Chun-Guey Wu, Henry O. Marcy, Donald C. DeGroot, Carl R. Kannewurf, Mercouri G. Kanatzidis, in Advanced Organic Solid State Materials (Mater. Res. Soc. Proc. 173, Pittsburgh, PA 1990) pp. 317-322.

6. Mercouri G. Kanatzidis, Chun-Guey Wu, Henry O. Marcy, Donald C. DeGroot, and Carl R. Kannewurf, Chem. Mater. 2, 222 (1990).

7. Y.-J. Liu, D C. DeGroot, J. L. Schindler, C. R. Kannewurf, and M. G. Kanatzidis, Chem. Mater. 3, 992 (1991).

8. J. Ross McDonald, Impedance Spectroscopy (John Wiley & Sons, New York, 1987).

9. J. Lemerle, L. Nejem, and J. Lebebvre, J. Inorg. Nucl. Chem. 42, 17 (1980).

10. J. Bullot, O. Gallais, M. Gauthier, and J. Livage, Appl. Phys. Lett. 36, 986 (1980).

11. C. Sanchez, F. Babonneau, R. Morineau, J. Livage, and J. Bullot, Philos. Mag. B 47, 279 (1983).

12. J. Bullot, P. Cordier, O. Gallais, M. Gauthier, and M. Livage, J. Non-Cryst. Solids 68, 123 (1984).

13. Y.-J. Liu and M. G. Kanatzidis (manuscript in preparation)

14. Joseph W. Lyding, Henry O. Marcy, Tobin J. Marks, and Carl R. Kannewurf, IEEE Trans. Instrum. Meas. 37, 76 (1988).

15. M. Pollak and T. H. Geballe, Phys. Rev. 122, 1742 (1961).

16. A. K. Jonscher, Nature 267, 673 (1977).

17. G. E. Pike, Phys. Rev. B 6, 1572 (1972).

PART III

Inorganic/Inorganic Composites

PART II

Inorganic Inorganic Compounds

MICROSTRUCTURE DEVELOPMENT AND MECHANICAL PROPERTIES OF Ce-TZP/La-β-ALUMINA COMPOSITES

TAKASHI FUJII*, HIRONOBU MURAGAKI*, HIRAKU HATANO* AND SHIN-ICHI HIRANO**
*Technical Research Center, Nisshin Flour Milling Co., Ltd.
5-3-1 Tsurugaoka, Oi-machi, Saitama 354, Japan
**School of Eng., Nagoya Univ.
Furo-cho, Chikusa-ku, Nagoya 464, Japan

ABSTRACT

Simultaneous additions of lanthanum aluminate(LAL) and Al_2O_3 to Ce-TZP (12mol% CeO_2-ZrO_2) lead to the in-situ formation of lanthanum-β-alumina(LBA) platelets ($\approx 1.0\,\mu m$ in width and $5 \sim 10\,\mu m$ in length) in the Ce-TZP matrix during sintering. The composites showed a fracture toughness(SEPB method) of 9.5 $MPa \cdot m^{0.5}$ and fracture strength of 960 MPa, which are remarkably improved from Ce-TZP sintered body (8.5 $MPa \cdot m^{0.5}$ and 560 MPa).
The composites also exhibit the no degradation by hydrothermal treatment.

INTRODUCTION

The application of zirconia-based ceramics has been extended to many aspects of advanced materials because of its good mechanical, thermal, chemical and electrical properties. Yttria-doped tetragonal zirconia polycrystals(Y-TZP) has received the great attention because of their excellent mechanical properties, wear properties and thermal expansion coefficient close to that of iron-based alloys. The high strength and fracture toughness of TZP are considered to be realized by the stress-induced transformation from tetragonal to monoclinic phase. However, it was reported that the mechanical properties are greatly degraded by low-temperature annealing in humid atmosphere or hot aqueous solution[1].
Ceria-doped tetragonal zirconia polycrystals(Ce-TZP) is superior in hydrothermal stability and fracture toughness. The strength of Ce-TZP is, however, relatively low and has been required to be improved by the control of microstructure.
Al_2O_3 doping is known to increase the fracture strength of Ce-TZP[2], but to decrease the fracture toughness. Whisker addition (such as SiC) is effective for improving the fracture toughness but also often give rise to the reduction of the fracture strength due to difficulty of the sintering.
On the other hand, Takahata et al. tried to improve the mechanical properties of alumina by growing needle-like crystals of lanthanum beta alumina(LBA) in the alumina matrix during course of sintering[3]. The authors reported the preliminary result of the effect of LBA formation in the Ce-TZP matrix on mechanical properties [4].
This paper describes the details of processing and microstructures of composites by sintering Ce-TZP, alumina and LBA in order to improve the fracture strength of Ce-TZP with high fracture toughness. The sintering condition, mechanical properties and hydrothermal stability of the composites were evaluated based on the microstructure development of LBA platelets in matrix.

EXPERIMENTAL PROCEDURES

Fabrication of LBA and Ce-TZP/LBA composites

The composition of LBA is $La_2O_3/11Al_2O_3$. Its structure resembles that of sodium beta alumina, which has been known as a popular solid electrolyte. Generally, materials with high anisotropic structure show crystal habits of large geometrical anisotropy.

LBA often takes an anisotropic shape as automorphism, also. Figure 1 shows the phase diagram in which there are two compounds with a ratio of 1:1 (perovskite structure) and 1:11 (beta alumina structure) in La_2O_3-Al_2O_3 system[5].

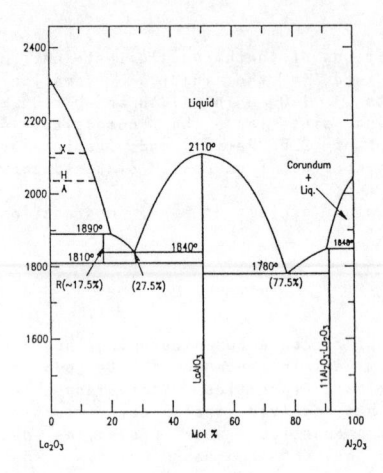

Figure 1 Phase diagram of La_2O_3-Al_2O_3 system [5].

The solid phase reaction between La_2O_3 and Al_2O_3 generates LBA through two stages as follows[6]. :

$$La_2O_3 + Al_2O_3 \rightarrow 2LaAlO_3 \quad (LAL)$$

$$LaAlO_3 + 5Al_2O_3 \rightarrow LaAl_{11}O_{18} \quad (LBA)$$

The raw powder materials were prepared by two processes shown above. Lanthanum oxalate and gamma-alumina were weighed, as the molar ratio of La_2O_3 and Al_2O_3 corresponds to 1:1. They were ball milled in ethanol, dried and calcined at 1100°C for 2 hours in air to synthesize $LaAlO_3$ powder.

The LBA powder was prepared from stoichiometric proportion of α-Al_2O_3 (Sumitomo Chem. : AKP-30) and $LaAlO_3$ prepared by the above processing.

To prepare the compound powder, 12 mol% Ce-TZP (Tosoh: TZ-12CE), α-Al_2O_3 and $LaAlO_3$ were added as weight ratio of 12Ce-TZP/Al_2O_3/LBA ; this formulation corresponds to 80:15:5 in weight percent. ;the mixture was then milled again in ethanol.

After drying , green compacts were fabricated first by die pressing the powder at 30 Mpa, followed by cold isostatic pressing(CIP) at 300 MPa without using any binders. The green compacts were sintered at 1600°C to 1625°C for 4 hours in air. Sintered billets were cut, ground and polished for characterization.

Microstructural characterization and mechanical properties

The microstructures of the sintered samples were characterized by scanning electron microscopy (Hitachi model S-800) using ground and polished specimens that were thermally etched for 15 minutes at 50°C below the sintering temperature. Relative phase contents (vol% of tetragonal and monoclinic phases) on as-sintered and fracture surfaces were determined by X-ray diffraction using the procedure described by Toraya et al. [7].

The fracture strength was evaluated by three-point bending (span: 30mm, cross-head speed: 0.5mm/min) on sintered samples cut into 3 × 4 × 45mm bars. The fracture toughness was evaluated using a single-edge-precracked-beam (SEPB) method[8]. A Vickers indentation-induced microcrack was introduced into a specimen, and then a sharp pop-in precrack was followed by applying the bridge indentation method. The sharply precracked specimen were fractured by three-point bending under a cross-head speed of 0.5mm/min. In these tests, fracture toughness was estimated from the fracture loads and precrack length.

Hydrothermal stability was evaluated by the hydrothermal treatment on the specimens cut for the bending test. The sample and distilled water were sealed in a Teflon-lined autoclave and then heated at 150°C for 120 hours. The hydrothermally treated specimen was checked by a three-point bending test.

RESULTS AND DISCUSSIONS

Mechanical properties of LBA

Figure 2 shows XRD patterns of as-sintered surface of LBA ceramics sintered at 1400°C to 1600°C in air. The starting powder was prepared with LaAlO$_3$ and α-Al$_2$O$_3$ as the composition corresponding to LBA. The XRD patterns of LBA ceramic sintered at 1400°C can be ascribed to LaAlO$_3$ and α-Al$_2$O$_3$. This result indicates that LBA is not formed at 1400°C yet. Referring to the (024) peak of LaAlO$_3$ which is never overlapped with peak of LBA, the amount of LaAlO$_3$ shows the tendency to decrease as sintering temperature was raised and diminished at 1600°C. The XRD patterns at 1600°C shows the presence of only LBA. Ropp et al. [5] reported that in case of the mixture of lanthanum oxide and aluminum oxide, only LaAlO$_3$ was formed at temperature 800°C to 1400°C and then LBA was formed above 1500°C.

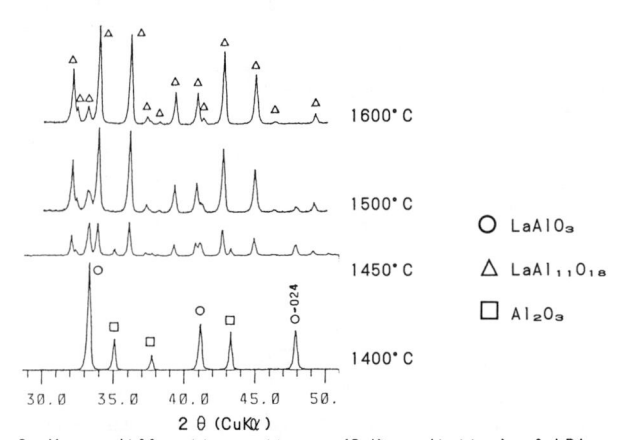

Figure 2 X-ray diffraction patterns (CuKα radiation) of LBA

○ LaAlO$_3$
△ LaAl$_{11}$O$_{18}$
□ Al$_2$O$_3$

Mechanical properties of LBA ceramics sintered at 1625°C are shown in Table I, indicating that LBA has similar value of the fracture strength, fracture toughness and thermal expansion coefficient to those of Al_2O_3 except for the Vickers hardness.

Table I Mechanical properties of LBA sintered at 1625°C for 4 hours.

	Fracture strength (MPa)	Fracture toughness (MPa·m$^{0.5}$)	Vickers hardness (GPa)	Thermal expansion coefficient ($\times 10^{-6}$/°C)
LBA	390 (355-440)	3.6	12.6	7.7

The SEM photograph of the thermally etched polished surface of a LBA ceramic in Figure 3 exhibits the formation of anisotropic platelet-like crystals of LBA.

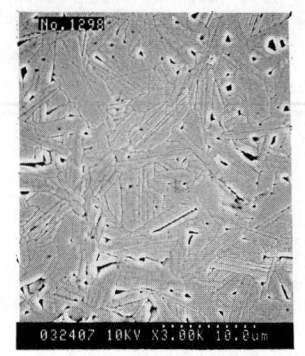

Figure 3 Microstructure of LBA sintered at 1625°C for 4 hours.

Figure 4 shows the temperature dependence of the fracture strength for LBA ceramics. The fracture strength did not decrease with rising temperature and it is noteworthy that LBA ceramic has a high strength value of about 370 MPa even at temperature as high as 1400°C.

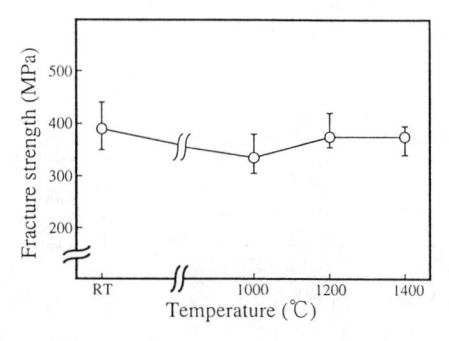

Figure 4 Temperature dependence of fracture strength of LBA sintered at 1625°C for 4 hours.

Microstructure and mechanical properties of Ce-TZP/LBA composite

Figures 5(A) and (B) show the microstructures of composites with Al_2O_3 or Al_2O_3/LBA at 1600°C for 4 hours. The platelet-shaped and grain shaped crystals are observed in the Ce-TZP matrix.

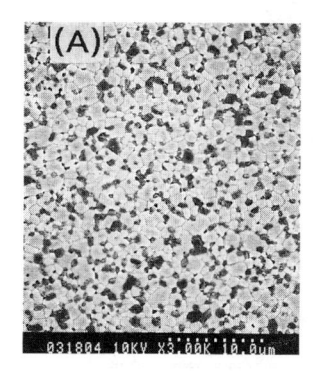

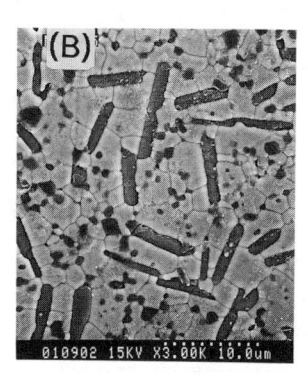

Figure 5 Microstructure of (A) 12Ce-TZP/Al_2O_3 and (B) 12Ce-TZP/Al_2O_3/LBA sintered at 1600°C.

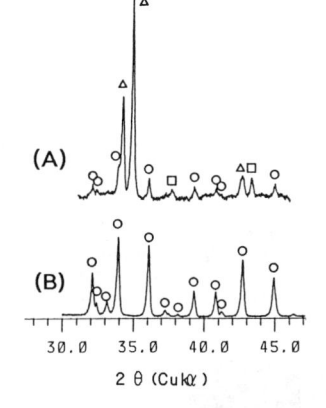

○ $LaAl_{11}O_{18}$ (LBA)
△ $t-ZrO_2$
□ Al_2O_3

Figure 6 XRD patterns of sintered surface of (A) composite and (B) LBA.

Figure 6 shows the XRD patterns of : (A) the surface of the composite sintered at 1600°C, and (B) the surface of LBA fabricated from $LaAlO_3$ and Al_2O_3 powder. These figures show peaks corresponding to the following phases : (A) LBA, ZrO_2, and Al_2O_3, and (B) LBA, respectively. This suggested that LBA was formed in situ during sintering, as would be expected based on phase diagram between $LaAlO_3$ and Al_2O_3 according to the reaction mentioned above.

Moreover, to observe the formation behavior of LBA, La and Al were mapped by energy dispersive X-ray analysis(EDX). This analysis demonstrates that platelets are correspond to LBA and the grains are Al_2O_3 particles.

The edge on SEM photograph of the sintered surface indicates that LBA grains are platelets (see Figure 7). The LBA platelets which formed in-situ were approximately 1 μm in thickness and 5 to 10 μm in length and width.

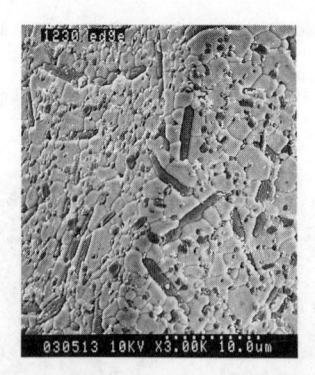

Figure 7 SEM photograph of 12Ce-TZP/Al₂O₃/LBA(the edge of the sintered surface).

Table II Mechanical properties of Ce-TZP base ceramics.

Al₂O₃ (wt.%)	La-β-Al₂O₃ (wt.%)	σf (MPa)	K_IC (MPa·m^0.5)	m-ZrO₂* (%)
0	0	560 (535-573)	8.5	73
20	0	820 (760-885)	6.5	60
15	5	960 (924-1010)	9.5	58

* Monoclinic ZrO₂/total ZrO₂ (%) on fracture surface.

As shown in Table II, the addition of Al₂O₃ to 12Ce-TZP increased the fracture strength but decreased the fracture toughness in agreement with the result of Tsukuma et al. [2].

Table II also shows data for 12Ce-TZP/AL₂O₃/LBA composite. The amount of monoclinic ZrO₂ on ground surfaces was approximately 2-5 % respectively. While the fracture toughness increased from 8.5 for 12Ce-TZP to 9.4 for composite with Al₂O₃ and LBA, the amount of monoclinic ZrO₂ on fracture surfaces decreased from 73 % for 12Ce-TZP to 58 % for LBA addition. These data suggest that the toughening mechanism by to transformation toughening is operative in these composite materials.

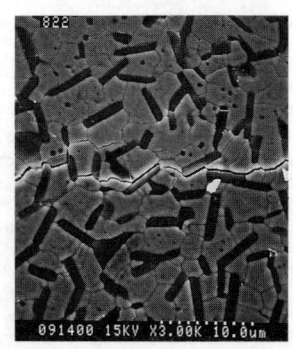

Figure 8 SEM photograph of crack propagation from Vickers indent of 12Ce-TZP/Al₂O₃/LBA ceramic.

A scanning electron micrograph of the crack propagation from Vickers indent of the composite is shown in Figure 8. The crack deflection by the platelets of LBA were clearly observed, which leads to the K_{IC} improvement in this composite as well as by the transformation of zirconia.

Hydrothermal stability of Ce-TZP/LBA composites

Table III shows the fracture strength and the volume percentage of an m-phase by XRD analysis, after hydrothermal treatment at 150°C for 120 hours : (a) 12Ce-TZP (b) 12Ce-TZP/Al$_2$O$_3$/LBA composite. According to the previous paper on Y-TZP[1], hydrothermal treatment gives rise to the increase in the amount of m-phase of zirconia at the surface of the body treated, and the reduction of the strength. Neither phase transformation nor a decrease in fracture strength after hydrothermal treatments were detected on both specimens of CeO$_2$-doped zirconia.

Table III Fracture strength and volume percentage of m-phase of 12Ce-TZP and 12Ce-TZP/Al$_2$O$_3$/LBA composite before and after hydrothermal treatment. (150°C, 120 hours)

	Fracture strength (MPa)		* vol.% of m-phase	
	before	after	before	after
12Ce-TZP	513 (490-535)	508 (501-543)	12.1	13.5
12Ce-TZP/Al$_2$O$_3$/LBA	910 (873-945)	905 (875-932)	4.0	5.0

*ground surface

CONCLUSIONS

(1) Platelets crystals of LBA grew homogeneously in-situ in the Ce-TZP matrix during sintering.
(2) High performance ceramic was obtained by sintering at 1600°C. Mechanical properties of the composite ceramic was 960 MPa in fracture strength and 9.5 MPa·m$^{0.5}$ in fracture toughness.
(3) Degradation of fracture strength was not detected by the hydrothermal treatment.

REFERENCES

1. T. Sato, T. Endo and M. Shimada, J. Am. Ceram. Soc. 72(5), 761-764 (1989).
2. K. Tsukuma, T. Takahata and M. Shiomi, Advance In Ceramics, Vol. 24, (The American Ceramic Society, Columbus, OH 1988), p. 721.
3. T. Takahata, K. Tsukuma and H. Yamamura, Yogyo-Kiso-Kagaku-Toronkai Proc. 3A03 (1987).
4. T. Fujii, H. Muragaki, H. Hatano and S. Hirano, Ceramic Transactions, Vol. 22, (The American Ceramic Society, Columbus, OH 1991), p. 693.
5. M. Mizuno, R. Berjoan, J. P. Coutures and M. Foex, Yogyo-Kyokai-shi, 92(12), 631-636 (1974).
6. R. C. Ropp and B. Carrol, J. Am. Ceram. Soc. 63(7-8), 416-19 (1980).
7. H. Toraya, M. Yoshimura and S. Somiya, J. Am. Ceram. Soc. 67(6), C-119-121 (1984).
8. T. Nose and T. Fujii, J. Am. Ceram. Soc. 71(5), 328-333 (1988).

Al$_2$O$_3$-ZrO$_2$ CERAMICS WITH SUBMICRON MICROSTRUCTURES OBTAINED THROUGH MICROWAVE SINTERING, PLASMA SINTERING AND SHOCK COMPACTION

J. McKITTRICK[*], B. TUNABOYLU[*], J.D. KATZ[**] AND W. NELLIS[***]
[*]University of California, San Diego, Materials Science Program, La Jolla, CA 92093
[**] Los Alamos National Laboratory, Los Alamos, NM 87545
[***]Lawrence Livermore National Laboratory, Livermore, CA 94550

ABSTRACT

Submicron and nanocrystalline grain sizes were achieved in the Al$_2$O$_3$-ZrO$_2$ eutectic composition through conventional, microwave and plasma sintering of rapidly solidified starting powders and through shock compaction of commercial powders. Post sintering studies revealed nanocrystalline intragranular ZrO$_2$ in the 1-2 μm Al$_2$O$_3$ grains, which is thought to be a result of the solidification synthesis. Additions of B$_2$O$_3$ greatly increased the final density through liquid phase sintering. Shock compression of commercial powders produced dense, crack-free, fine grained ceramics with loading pressures up to 9.1 GPa and a metastable ZrO$_2$ phase under higher pressures.

INTRODUCTION

Rapid solidification processing (RSP) is a technique used to generate materials with unusual microstructures and properties which may not be obtained through other methods. With increasing solidification rates, submicrocrystalline, nanocrystalline and/or amorphous structures may be quenched into the solid from the melt. Al$_2$O$_3$-ZrO$_2$ composites are of interest for various structural applications [1-3]. Finer grain sizes in this system result in higher fracture toughness [4-6] and strength [7-9]. The metastable structures of the RSP'd materials may not necessarily be retained after conventional densification, which typically involves sintering at high temperatures. Microwave and plasma heating are sintering techniques in which the samples are heated rapidly to high temperatures to promote densification but inhibit grain growth [10,11].

Shock compaction of ceramic powders is a technique which can (a) form dense compacts in short times, (b) provide physical and chemical bonding at the powder interfaces and (c) create mechanochemical effects [12]. The mechanochemical effects include such shock processing characteristics such as high defect densities, melting and rapid solidification at the powder interfaces and high pressure phase transformations. Compositions in the Al$_2$O$_3$-ZrO$_2$ system require high temperature sintering to densify and are ideal materials to examine as possible candidates for this low temperature compaction.

Our goal was to examine the microstructural development of several non-conventional compaction techniques as applied to the Al$_2$O$_3$-ZrO$_2$ system and evaluate the effectiveness of the techniques for producing dense, homogeneous bodies.

EXPERIMENTAL TECHNIQUES

The eutectic composition was used for all studies which has a melting temperature of ~1800°C. Technical grade commercial powders of 61.95 mol % (57.4 wt%) α-Al$_2$O$_3$ and 38.05 mol% (42.6 wt%) monoclinic ZrO$_2$ (m-ZrO$_2$) were mixed and pressed to form rods of ~6 cm length and 1 cm in diameter. The rods were suspended above a twin roller solidification device and were heated with an oxy-hydrogen torch. Liquid droplets formed on the end of the rod and were allowed to drop between the rollers rotating with surface speeds of ~10 m/sec.

The as-quenched materials were ball milled in ethanol overnight with Al$_2$O$_3$ or B$_4$C media, dried and then sieved through 45 μm. The resultant starting powder was composed of angular and sharp particles with a broad size distribution. The powders were cold pressed under 350 MPa into 1.27 cm diameter, 4 mm thick disks and then sintered for 2 hours at

1400°C or 1600°C with a ramp of 10°C/min. Other samples were microwave sintered at 1400°C or 1600°C. The average ramp time to the high temperature was 10-15 min. The samples were not held at T_{max}. The microwave furnace consists of a 2.45 GHz, 6 kW generator and a 5.7 x 10^{-2} m^2 resonant cavity. Plasma sintering[¶] was done at 1500 or 1550°C for 2 min.

For shock compaction studies, the commercial powders were additionally sized with electroformed sieves through 20 μm. The powders were loaded into the gas gun facility at Lawrence Livermore National Laboratory and then subjected to pressures ranging from 6.3 to 41 GPa.

All of the compacted samples were examined by X-ray diffraction, electron microscopy and the bulk densities were measured using the Archimedes' principle.

EXPERIMENTAL RESULTS AND DISCUSSION

Sintering Studies

The as-quenched samples were nanocrystalline or sub-microcrystalline α-Al_2O_3 and t-ZrO_2, as seen by X-ray diffraction. X-ray line broadening measurements indicate that the crystalline size is on the order of 30-100 nm. This is in agreement with other results found in these materials by previous work [13] and other workers [14].

Microstructural differences between materials milled with Al_2O_3 and B_4C during conventional sintering is seen in Fig. 1 (a) and (b). Both materials were heated to 1200°C for 2 hours. The material milled in the Al_2O_3 shows no features and the coarse and angular nature of the starting material is clearly seen. No sintering has occurred. For the material milled in the B_4C, small nucleation events are observed in the starting material. Small nuclei which range from 200-800 nm occur homogeneously on the surface. This difference can be attributed to the boron contamination in the powder. The phase diagram for Al_2O_3-B_2O_3 [15] indicates that a B_2O_3-rich liquid will be present at 1200°C along with a $9Al_2O_3$-$2B_2O_3$ solid phase if ~18 mol% B_2O_3 is present. The orthorhombic $9Al_2O_3$-$2B_2O_3$ was found in the X-ray diffraction trace. As the material was heated to 1400°C, 'rods' of the $9Al_2O_3$-$2B_2O_3$ phase are found dispersed with small, round grains of ZrO_2, as shown in Fig 1 (c). The rods are several microns long and about 200 nm in diameter. Rod-type growth is often found in systems in which a small amount of liquid is present. At 1600°C, the microstructure consists of small ZrO_2 grains (light phase) ranging from 200 nm - 1 μm in diameter dispersed in the Al_2O_3 matrix (dark phase), as shown in Figure 1 (d).

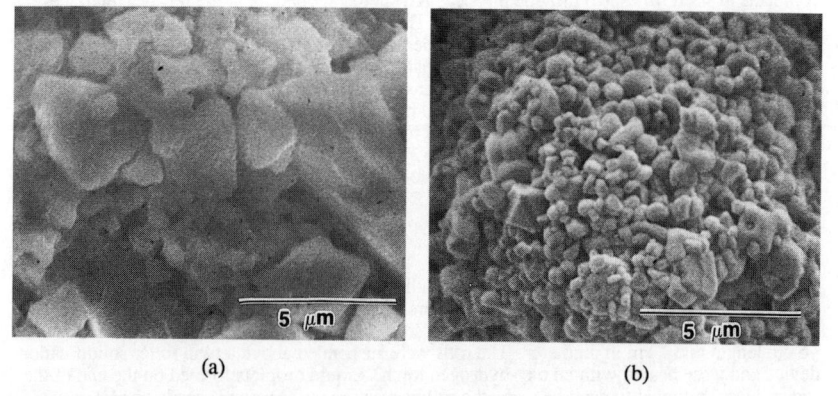

(a)   (b)

Figure 1. SEM micrographs of conventionally sintered material. (a) Milled with Al_2O_3 media, 1200°C, 2 hours (b) Milled with B_4C media, 1200°C, 2 hours

---

[¶] Sintering done at the New Mexico Institute of Mining and Technology by Murat Bengisu.

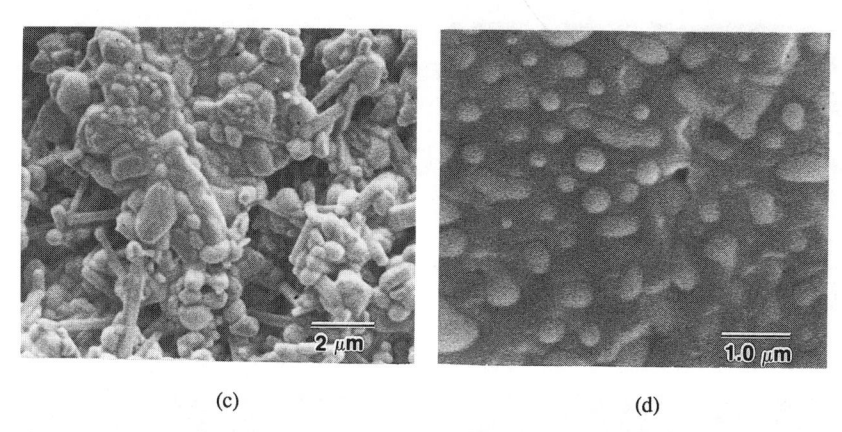

Figure 1. (cont.) SEM micrographs of conventionally sintered material. (c) Milled with B$_4$C media, 1400°C, 2 hours. 'Rod' morphology is the 9Al$_2$O$_3$-2B$_2$O$_3$ phase (d) Milled with B$_4$C media, 1600°C, 2 hours

Microwave sintering shows that interparticle sintering occurs, as shown in Fig. 2 (a). The high degree of surface porosity can be attributed to the broad size distribution of the starting material. Higher magnifications shown in Fig. 2 (b) and (c) show 1-2 μm Al$_2$O$_3$ grains with intra- and intergranular ZrO$_2$ grains. The intragranular ZrO$_2$ has a very fine grain size, ranging from 100-200 nm whereas the intergranular ZrO$_2$ is coarser, on the order of 0.5-1 μm. The intragranular ZrO$_2$ is thought to be a result of the solidification synthesis. It is well known that rapid solidification produces supersaturation of one component in the other. High degrees of supersaturated ZrO$_2$ in the Al$_2$O$_3$ would precipitate in the grains at higher temperatures. These precipitates would be inhibited from growing further due to long range diffusional problems as well as physical confinement in a well-crystallized region.

Plasma sintering resulted in denser samples, presumably due to cold isostatic pressing the powders prior to sintering, as shown in Fig. 3 (a). At higher magnifications in Fig 3 (b), a similar microstructural morphology of the two phases as that of microwave sintering is seen.

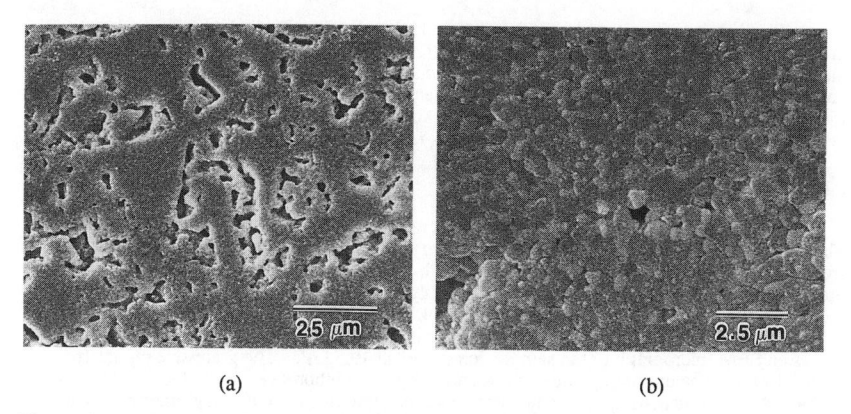

Figure 2. SEM micrographs of microwave sintered material. The material was milled with B$_4$C media (a) low magnification showing surface porosity (b) higher magnification showing interparticle sintering

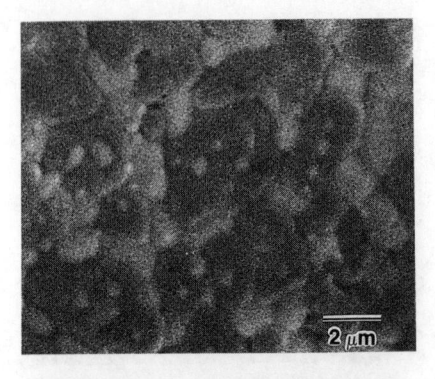

Figure 2. (cont.) SEM micrographs of microwave sintered material. The material was milled with B$_4$C media (c) intra- and intergranular ZrO$_2$

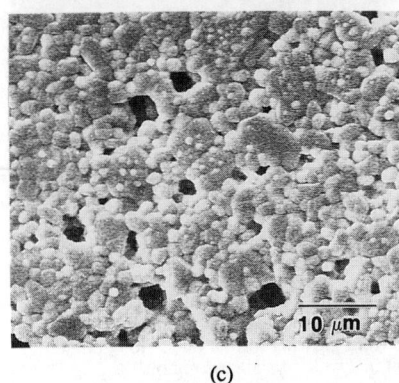

(c)   (d)

Figure 3. Plasma sintered material. (a) low magnification showing reduced surface porosity (b) higher magnification showing similar microstructure to that of microwave sintering.

The boron contaminated material showed higher densities for both sintering procedures, presumably due to liquid phase sintering with the B$_2$O$_3$-rich liquid. As expected, the microwave sintered material had the highest density observed: 95.8% ρ_{th} in the boron contaminated material sintered at 1600°C. The densities of the sintered samples were higher for the B$_4$C milled material and with microwave sintering. The densities range from 65-93% ρ_{th} for the Al$_2$O$_3$ milled material and 70-96% ρ_{th} for the B$_4$C milled material.

Shock Compaction

Fig. 4 shows an optical micrograph of a crack-free sample compacted under 6.3 GPa. The density of this sample was 88% ρ_{th}. Bengisu et al. [16] have reported fabricating a crack-free sample with 5.7-8.7 GPa but they report high porosities ranging from 30-40%. Higher pressures produced macrocracked samples with higher densities. Eventually only powder was recovered at 41 GPa, the highest pressure applied.

The samples retained a fine grained microstructure under the shock treatment. Fig. 5 shows an SEM micrograph of the sample compacted at 19.3 GPa. The grain size ranges from 0.16-1.15 µm. The micrograph does show some degree in inhomogeneity in the separation of the two phases. Our starting powders were not completely free of Al$_2$O$_3$ agglomerates. The SEM indicates that the agglomerates were sheared but remained in local proximity to each other. At higher pressures, the ZrO$_2$ began to transform to the tetragonal or high pressure orthorhombic phase, as shown in Fig. 6. From the data, we could not determine which metastable phase was present.

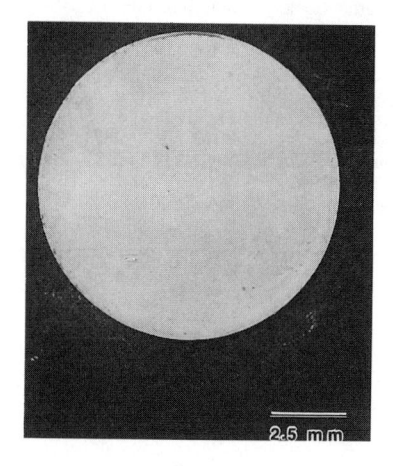

Figure 4. Optical micrograph of a crack-free sample compacted at 6.3 GPa.

Figure 5. SEM micrograph of the material shock compressed at 6.3 GPa.

SUMMARY

Compaction of rapidly solidified Al_2O_3-ZrO_2 eutectic through conventional, microwave and plasma sintering produces composites composed of nanocrystalline ZrO_2 and microcrystalline Al_2O_3. Additions of boron produce a whisker (Al_2O_3-rich) reinforced composite of higher density, due to liquid phase sintering. The bulk densities for the sintering procedures could be optimized by obtaining a finer particle size distribution of the milled, as-quenched material. Shock compaction produces dense, crack-free, fine grained materials at pressures < 9.1 GPa. At higher pressures, microcracks form and the monoclinic ZrO_2 transforms to either a tetragonal or orthorhombic polymorph.

ACKNOWLEDGEMENTS

This work was supported by the Advanced Technology Assessment Center at Los Alamos National Laboratory and the Institute of Geophysics and Planetary Physics at Lawrence Livermore National Laboratory.

REFERENCES

[1] R.C. Garvie, R.H. Hannink and R.T. Pascoe, *Nature*, **258** 703-704 (1975)
[2] U. Dworak, H. Olapinski, D. Fingerli and U. Krohn, Adv. in Ceram., eds. N. Claussen, M. Rühle and A. H. Heuer, The American Ceramic Society, Columbus, OH. **12** 480-487 (1983)
[3] L. Coes, Abrasives, (Springer-Verlag, New York, 1971), p. 61-67
[4] N. Claussen, *J. Am. Ceram. Soc*, **59** 49-51 (1976)
[5] F. F. Lange, *J.Mater.Sci.*, **17** 247-254 (1982)
[6] D.J. Green, *J.Am.Ceram.Soc.*, **65** 610-614 (1982)
[7] K. Tsukuma, K. Ueda and M. Shimada, *J.Am.Ceram.Soc.*, **68** C4-C5 (1985)
[8] S. Hori, R. Kurita, M. Yoshimura and S. Somiya, Adv. in Ceram., eds. S. Somiya, N. Yamamoto and H. Yanagida, The American Ceramic Society, Columbus, OH. **24A** 423-429 (1986)
[9] M. Rühle, A. Strecker , D. Waidelich and B.Kraus, Adv. in Ceram., eds. N. Claussen,

M. Rühle and A. H. Heuer, The American Ceramic Society, Columbus, OH. **12** 256-274 (1983)
[10] J. D. Katz and R. D. Blake, *Am.Ceram.Soc.Bull*, **70** [8] 1304-1308 (1991)
[11] W.H. Sutton, *Am.Ceram.Soc.Bull*, **68** [8] 376-86 (1984)
[12] A.B. Sawaoka, Ceramic Industrial Processing, eds. L.E. Murr, K.P Staudhammer and M.A. Meyers (Marcel Dekker, Inc. Publishers, NY 1986) p. 221-229
[13] J. McKittrick, G. Kalonji and T. Ando, *J.Non-Cryst.Sol..*, **94** 163-171 (1987)
[14] V. Jayaram, C. Levi, T. Whitney and R. Mehrabian, *Mat.Sci.Eng.*, **A124** 65-81 (1990)
[15] P.J.M. Gielisse and W.R. Foster, *Nature*, **195** 70 (1962)
[16] M. Bengisu, O. Inal and J. Hellman, *J. Am. Ceram. Soc.*, **73** 346 (1990)

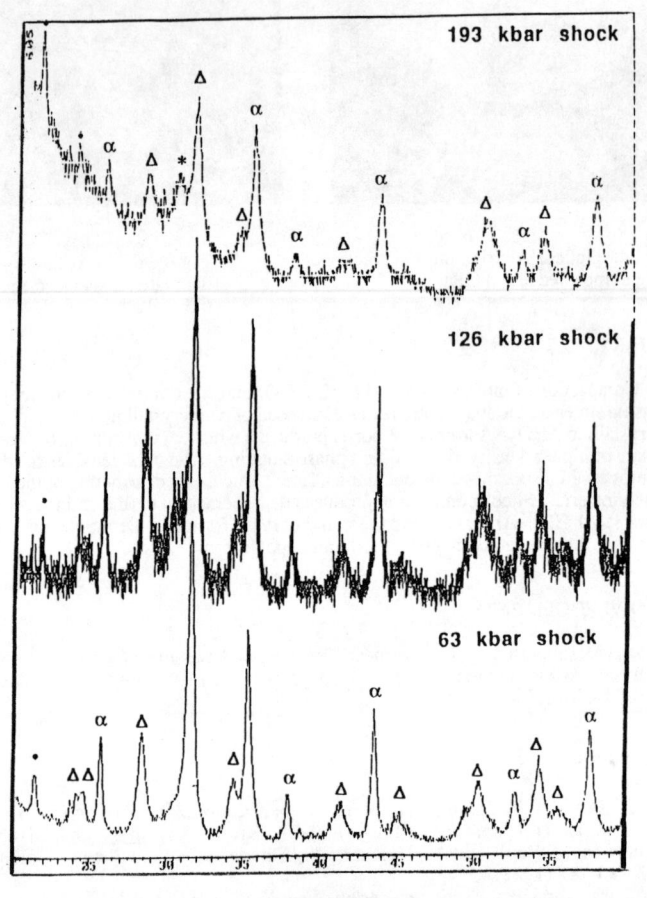

Figure 6. X-ray diffraction traces taken on the material shocked compresses at 63, 126 and 193 kbar (6.3-19.3 GPa). Note shoulder forming on 100% monoclinic ZrO_2 peak at the high pressures. This is either tetragonal or orthorhombic ZrO_2. α = α-Al_2O_3, Δ = monoclinic ZrO_2, * = orthorhombic or tetragonal ZrO_2, • = unidentified

MICROCRYSTALLINE CERAMIC COMPOSITES BY ACTIVE FILLER CONTROLLED REACTION PYROLYSIS OF POLYMERS

PETER GREIL, MICHAEL SEIBOLD AND TOBIAS ERNY
Technical University of Hamburg–Harburg, Advanced Ceramics Group, Denickestr. 15, 2100 Hamburg 90, Germany

ABSTRACT

Pyrolytic conversion of preceramic polymers such as polysilanes, –silazanes, or –siloxanes to ceramics may be significantly influenced by the presence of active filler dispersoids. Based on thermodynamic and microstructural considerations a variety of suitable polymer–filler systems can be found which allow the fabrication of microcrystalline composite materials with low dimensional change upon polymer–ceramic conversion. As an example the active filler controlled reaction pyrolysis of polysiloxane with addition of titanium powder was investigated. A composite material with microcrystalline titanium carbide inclusions embedded in an amorphous (< 1000 ºC) or nanocrystalline (>1000 ºC) silicon oxycarbide matrix was formed. Property changes with increasing pyrolysis temperature can be attributed to various microstructural transformations. Thus, a variety of potential fillers may be used to tailor the microstructure of polymer–derived ceramic composite materials in order to fabricate bulk materials and components with a broad range of compositions and properties.

INTRODUCTION

Manufacturing of ceramic materials in the system Si–C–N–O from preceramic silicon containing polymers such as polysilanes, –carbosilanes, –silazanes, or –siloxanes has recently attained particular interest [1]. Tailoring of molecular structure and composition of the appropriate starting compound is highly attractive because of its potential to optimize the processing behavior and to fabricate new materials with unique microstructure, such as nanostructured and molecular composites. Low fabrication temperatures < 1500 ºC and versatile plastic shaping technologies favor the polymer route for the fabrication of fiber reinforced composite ceramics and components with complex geometry. Although a variety of silicon containing polymeric precursors with high ceramic yields exceeding 75 wt.% could be developed [2], only low–dimensional products such as ceramic surface coatings or fibers have achieved greater significance [3].

The major limitations for bulk component fabrication from polymeric precursor systems are related to a tremendous shrinkage and density increase encountered in the pyrolytic polymer–ceramic conversion. For example, density typically increases by a factor of two to three from the precursor ($\rho \approx 1$ g/cm^3) to the ceramic residue (SiO_2 : $\rho \approx 2.2 - 2.6$; Si_3N_4, SiC : $\rho \approx 3.0 - 3.2$ g/cm^3) and linear shrinkages of more than 30 % can occur which usually result in extended cracking and pore formation in the pyrolyzed products [4]. During the pyrolysis of polycarbosilanes, for example, a variety of gaseous decomposition products like CO, H_2, hydrocarbons (CH_4, C_2H_6, C_6H_6 etc.) and silanes ((CH_3)$_4$Si, (CH_3)$_3$SiH, (CH_3)$_2$SiH$_2$ etc.) could be observed which have to be removed from the precursor material by diffusional transport through a network of open pore channels [5].

The formation of bulk components from organosilicon polymer/active filler mixtures has been reported as a novel processing route for the fabrication of oxycarbide ceramic composite materials [6,7]. In the presence of a carbide, silicide, or nitride forming transition metal (Ti, Cr, V, ...) the dimensional change associated with the polymer pyrolysis could be drastically reduced, and hence facilitated production of bulk components with controlled shrinkage and porosity.

For example, SiOC/TiC and SiOC/Cr$_3$C$_2$,SiC composite materials were prepared from polysiloxane/titanium– and polysiloxane/chromium disilicide–powder mixtures in Ar–atmosphere at temperatures below 1400 ºC. Recently, polysilazanes and –carbosilanes were demonstrated to be suitable reactants for the active filler controlled polymer pyrolysis process [8].

Suitable preceramic polymers may be chosen from a variety of available ceramic precursors, where those with high ceramic yields above 75 wt.% seem to be favourable. Oxygen containing polysiloxanes with Si–O–backbone structure provide a simple and inexpensive way for the formation of Si–C–O–matrices [9] for use in the intermediate (800–1200 ºC) temperature range, whereas polysilazanes and –carbosilanes with low oxygen content are of special interest for Si–N–C–matrices in composite materials to be used at higher temperatures.

BASIC PRINCIPLE

In the presence of an active filler Me (Me = transition metal) the polymer pyrolysis reaction is significantly changed due to chemical and physical interactions of the filler powder with constituents of the polymer phase. The overall chemical reaction may be expressed by

$$\left[-\underset{|}{\overset{R}{\underset{|}{Si}}}-X-\right] + Me \longrightarrow Si-X(O) + Me-C + RH_z + H_2$$

polymer + filler $\longrightarrow$ matrix + dispersant + volatiles

precursor mixture $\longrightarrow$ ceramic composite

where X is C (polycarbosilane), N (polysilazane) or O (polysiloxane), and R denotes carbon containing functional substituents like alkyl, allyl, vinyl, etc. groups.

In the case of a polysiloxane, [Si–O]$_n$, a silicon oxycarbide (Si–O–C) glass is formed upon polymer decomposition from which C, SiC and SiO$_2$ crystallize as the major phases above 1000 ºC [10]. Hence, some of the simplified possible overall reactions of an active filler Me would be

$$Me + xC + SiO_2 \longrightarrow MeC_x + SiO_2 \quad (1)$$

$$Me + (3-x)C + SiO_2 \longrightarrow MeO_x + SiC + (2-x)CO \quad (2)$$

$$Me + 2xC + xSiO_2 \longrightarrow MeSi_x + 2xCO \quad (3)$$

which result in the formation of binary carbides, oxides, or silicides of various compositions, with additional release of gaseous CO. Although additional ternary oxycarbides and carbosilicides may be formed as stable phases or metastable reaction layers at the filler/matrix interface, thermodynamic stability criteria for specific filler reaction products can be derived from reactions (1) – (3), at least in a first approximation [11]. Fig. 1 shows the relative stability $\Delta G_1 - \Delta G_2$ vs. $\Delta G_1 - \Delta G_3$ (Gibbs free energy data from [12]) of a variety of transition metal carbides against oxidation or silicide formation at 1000 ºC. Due to their extremely high melting points, excellent high temperature strength, good corrosion resistance, high hardness and electrical properties [13], a number of transition metal carbides such as TiC, NbC, Cr$_3$C$_2$, TaC and WC seem to be of particular interest for use as reinforcing compounds in carbide and oxycarbide ceramic matrices.

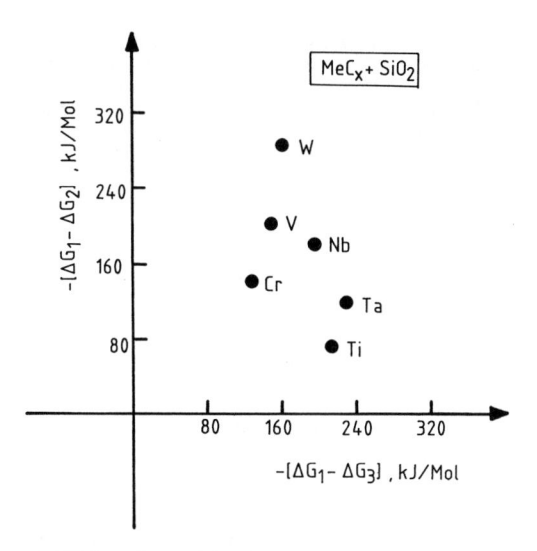

Fig. 1 Relative stabilities of transition metal carbides in a polysiloxane derived matrix phase at 1000 °C according to eqns. (1) – (3).

In the presence of a reactive atmosphere such as N_2, NH_3, BH_3 etc. the filler phase may react to form nitrides or borides, resulting in a weight increase instead of a weight loss as in inert atmosphere [11]. Thus, variation of polymer precursor, active filler phase, and reactive atmosphere allows tailoring of polymer–derived materials with controlled porosity, low shrinkage and superior performance, Fig. 2.

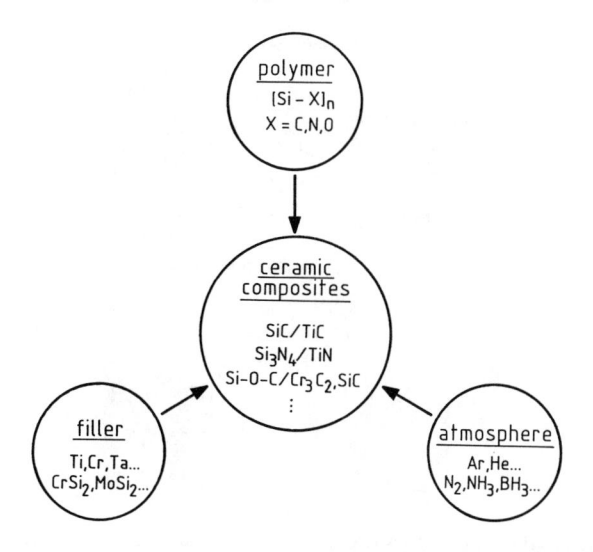

Fig.2 Tailoring ceramic compositions by variation of polymer, filler and atmosphere compositions in the active filler controlled polymer pyrolyis process.

The reduction of shrinkage and porosity generation during polymer–ceramic conversion is considered as a key aspect for the use of polymer pyrolysis technology in bulk component fabrication [14]. The pyrolytic degradation of a preceramic polymer precursor (P) results in a series of complex thermally induced processes, such as rearrangement, crosslinking and cleavage of carbon–hydrogen bonds, which yield condensed ceramic (C) and gaseous (G) reaction products

$$P(s,l) \xrightarrow{\Delta} C(s) + G(g) \qquad (4)$$

During pyrolysis of the polymer body, its mass is reduced by loss of volatile species, m(G), and the ceramic yield may be expressed by the ratio of mass of ceramic residue, m(C), to initial polymer mass, m(P),

$$\alpha = \frac{m(C)}{m(P)} = 1 - \frac{m(G)}{m(P)}. \qquad (5)$$

Due to the higher densities of the ceramic product(s), $\rho(C)$, compared to the starting polymer, $\rho(P)$, a density ratio may be defined

$$\beta = \frac{\rho(P)}{\rho(C)}. \qquad (6)$$

While for unconstrained microstructural relaxation, transient porosity is assumed to be eliminated by viscous or diffusional material transport at higher temperatures, and the maximum volume change during polymer to ceramic conversion is given as

$$\psi = \alpha\beta - 1, \qquad (7)$$

volume invariant conversion ($\psi = 0$) should result in a maximum porosity, π, of

$$\pi = 1 - \alpha\beta. \qquad (8)$$

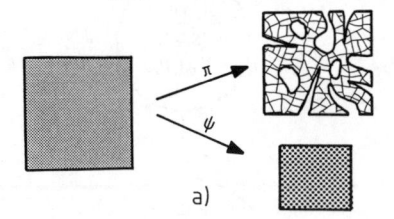

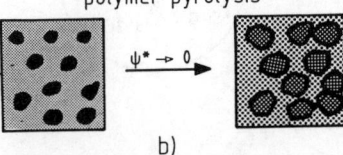

Fig.3 Microstructural changes encountered in polymer–ceramic conversion: a) filler–free polymer pyrolysis involving extended porosity, π, or shrinkage, ψ, and b) active filler controlled polymer pyrolysis with near net shape precursor–ceramic conversion.

Fig. 3a schematically shows the two extreme cases of polymer–ceramic conversion. Usually, a significant volume change as well as massive porosity formation are observed in real systems. When the product $\alpha\beta$ becomes smaller than unity, as it is the case in all organosilicon polymer precursors, Table I, it appears to be impossible to fabricate fully dense ceramic products from a polymer precursor without shrinkage.

Table I. Pyrolysis properties or organometallic polymers used as pre-ceramic precursor materials [14].

Polymer precursor	Products	Temperature (°C)/atmosphere	α	β
Hydridopolysilazane	Si_3N_4	1200/N_2	0.74	0.44
Vinylphenylpolysilazane	Si_3N_4	1000/N_2	0.85	
Polyborasilazane	BN/Si_3N_4	1000/Ar	0.90	0.55
Polycarbosilane + Al amide	Si_3N_4/AlN	1000/NH_3	0.54	
Polycarbosilane + Al amide	SiC/AlN	1000/NH_3	0.4	
Cyclomethylpolysilazane	Si_3N_4/SiC	1000/Ar	0.88	
Methylvinylpolysilane	SiC	1000/	0.83	
Polycarbosilane + Ti butoxide	SiC/TiC	1400/N_2	0.72	
Methylpolysiloxane	SiO_2/SiC	1000/He	0.85	0.46
Diphenylpolyborosiloxane	SiC/B_4C	950/Ar	0.45	0.58
Aminoborazine	BN	1000/Ar	0.55	
Polyiminoalane	AlN	600/N_2	0.42	
Ti propoxide ethanolamine	TiN	1600/N_2	0.85	

A substantial change of the situation is obtained by using an active filler phase (T) which can react with solid or gaseous decomposition products of the polymer, e.g. carbon (K) or hydrocarbon species (G), to form a new carbide phase (M)

$$P(s,c) + T(s) \xrightarrow{\Delta} C(s) + M(s) + G^i(g). \quad (9)$$

The total maximum volume change of the active filler containing precursor, ψ^*, may then be expressed by [14]

$$\psi^* = (1 - V_T/V_T^*)(\alpha\beta - 1) + V_T(\alpha^{TM}\beta^{TM} - 1). \quad (10)$$

V_T^* is a critical filler volume fraction in the starting mixture

$$V_T^* = V_T^{max}(3 - \alpha\beta - \alpha^{TM}\beta^{TM}) - (1 - \alpha\beta) \quad (11)$$

which determines the maximum particle packing density of the reacted filler phase in the pyrolyzed product. V_T^{max} is the maximum filler particle packing density (0.74 for fcc or hcp of equal sized spheres, and < 0.5 for randomly packed spheres of unequal size) and α^{TM} and β^{TM} describe the mass change of the filler phase

$$\alpha^{TM} = \frac{m(M)}{m(T) + m(K)} \quad (12)$$

and the density ratio

$$\beta^{TM} = \frac{\rho(T+K)}{\rho(M)}. \quad (13)$$

For $\alpha^{TM}\beta^{TM} > 1$ a volume expansion of the filler phase may compensate for the polymer shrinkage, Fig. 3b. Table II summarizes some of the potential filler systems and their characteristic volume changes.

Table II. Specific volume changes upon reaction pyrolysis of potential fillers [14].

Filler	$\alpha^{TM}\beta^{TM}$ Carburization (solid)	Carburization (gaseous)	Nitridation (gaseous)
Ti	0.76 (TiC)	1.14 (TiC)	1.08 (TiN)
V	0.79 (VC)	1.28 (VC)	1.27 (VN)
Cr	0.83 (Cr$_3$C$_2$)	1.25 (Cr$_3$C$_2$)	1.50 (CrN)
Zr	0.79 (ZrC)	1.09 (ZrC)	1.03 (ZrN)
Nb	0.85 (NbC)	1.27 (NbC)	1.35 (NbN)
Mo	0.95 (Mo$_2$C)	1.22 (Mo$_2$C)	—
Hf	0.83 (HfC)	1.17 (HfC)	1.04 (HfN)
Ta	0.86 (TaC)	1.27 (TaC)	1.25 (TaN)
W	0.84 (WC)	1.32 (WC)	—
Al	1.09 (Al$_4$C$_3$)	1.53 (Al$_4$C$_3$)	1.26 (AlN)
B	0.93 (B$_4$C)	1.20 (B$_4$C)	2.42 (BN)
Si	0.70 (SiC)	1.07 (SiC)	1.13 (Si$_3$N$_4$)
Fe	0.88 (Fe$_3$C)	1.10 (Fe$_3$C)	1.39 (FeN)

Assuming isotropic dimensional changes, linear shrinkage ϵ is obtained from volume shrinkage ψ by

$$\epsilon = (\psi + 1)^{1/3} - 1 \qquad (14)$$

with ϵ and ψ corresponding to the polymer and ϵ^* and ψ^* to the polymer/filler-system, respectively. In Fig. 4 the normalized linear shrinkage ϵ^*/ϵ is plotted versus the normalized filler volume fraction V_T/V_T^*. With increasing filler expansion characteristics ($\alpha^{TM}\beta^{TM} > 1$), the total linear shrinkage decreases for a given filler volume fraction in the starting mixture. The filler volume fraction to obtain zero shrinkage, where polymer shrinkage is fully compensated by the filler expansion, decreases from $V_T/V_T^* = 1$ for $\alpha^{TM}\beta^{TM} = 1$ to approximately 0.6 for $\alpha^{TM}\beta^{TM} = 2$.

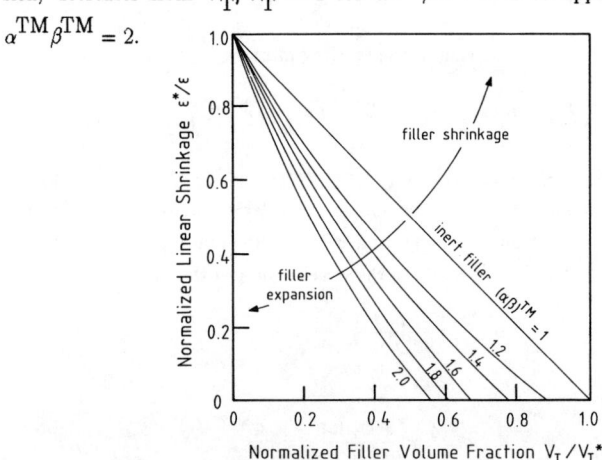

Fig. 4 Normalized linear shrinkage of polymer/filler-systems as a function of normalized filler volume fraction in the starting precursor mixture ($V_T^* = 0.5$, $\psi = -0.6$).

EXPERIMENTAL RESULTS IN THE SYSTEM POLYSILOXANE/TITANIUM

Based on the thermodynamic and geometrical considerations the system polysiloxane/titanium was chosen to investigate the microstructural changes which occur during filler controlled reaction pyrolysis in Ar–atmosphere. A commercially available liquid polysilsesquioxane (H 62 C, Wacker Chemie, Burghausen, GE) with an approximative composition of $[RSiO_{1.5}]_n$, with $R = C_6H_5$ as the major constituent and minor contents of CH_3, $CH_2=CH$ and H (molar ratio 2.8 : 1.5 : 1 : 1), was homogeneously mixed with various amounts of a Ti–powder (D 0681, Ventron alfa chemicals, Karlsruhe, GE), with a mean grain size of 1–3 μm. Specimens were prepared by pressurelessly casting the suspension into brass molds, followed by subsequent heating to 150 – 200 ºC in order to stabilize the green compact by a hydrosilation crosslinking reaction. The pyrolysis experiments were carried out in an electrically heated tube furnace at temperatures between 400 and 1400 ºC. Linear shrinkage and porosity were measured by standard procedures of differential dilatometry and pycnometry, respectively. Microstructure was analyzed by XRD and TEM techniques. Details of the specimen processing were described elsewhere [10].

Shrinkage and porosity

Fig. 5 shows the total linear shrinkage ϵ^* of the polysiloxane/titanium precursor mixture measured after pyrolysis in Ar–atmosphere at 1200 ºC for 4 h where polymer decomposition as well as filler carburization reactions are completed [10]. For comparison linear shrinkage was calculated according to eqns. (10) and (14) using $V_T^* = 0.5$ and $\alpha\beta = 0.4$. For the filler carburization the two cases assuming reaction with solid carbon, $\alpha^{TM}\beta^{TM} = 0.76$, or gaseous (hydro)carbon, $\alpha^{TM}\beta^{TM} = 1.14$, may define possible minimum and maximum values of shrinkage. Actually, the calculations indicate that a mixed mode of carburization reaction involving both solid and gaseous reactants should take place.

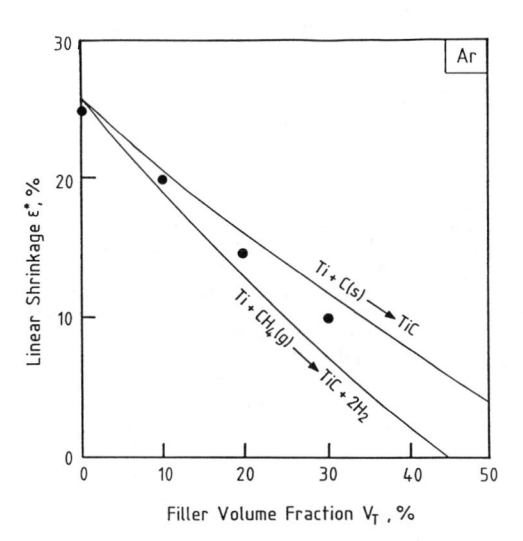

Fig. 5 Total linear shrinkage ϵ^* of polysiloxane filled with titanium powder measured at 1200 ºC (●) and calculated according to eqns. (10) and (14).

As a result of the structural rearrangements during thermal decomposition of the polymer and filler carburization, the density of the reaction product exhibits a systematic increase with pyrolysis temperature, Fig. 6. Calculated densities using a rule of mixture based on volume additives for compounds in their appropriate states (amorphous or crystalline) correlate well with the experimental data except for the temperature range of 600 – 800 ºC where the volatile decomposition products have to be released from the decomposing polymer. SEM analysis revealed a network of a transient pore channel system with diameters of 100 – 300 nm after pyrolysis at 600 ºC [10]. Similar transient pore systems have also been postulated to occur in polysilazane or –carbosilane pyrolysis products at intermediate temperature ranges [5] which will be removed by viscous flow or sintering at higher temperatures.

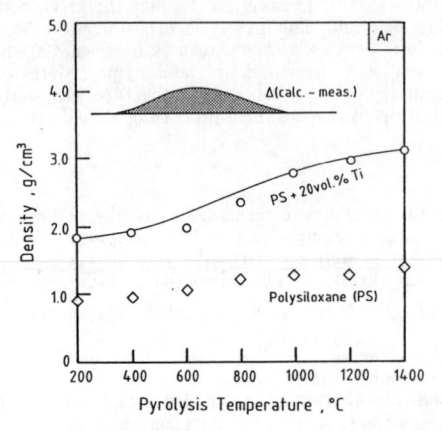

Fig. 6 Density vs. pyrolysis temperature as a function of the phase composition in the reaction product of a polysiloxane/20 vol.% titanium powder mixture (o,◊ measured, ——— calculated).

Microstructure

The shrinkage and density increase observed during heating the polymer/filler mixture result from various microstructural transformations which are associated with polymer decomposition and simultaneous filler carburization. During heating, decomposition of the polysiloxane polymer starts above 200 ºC and is almost completed at approximately 800 ºC. Above this temperature no further weight loss was observed indicating an inorganic nature of the Si–O–C (silicon oxycarbide) glass. Nanoparticles of excessive carbon are embedded in this amorphous phase which have a well–developed two–dimensional turbostratic structure similar to pyrolytic carbon [15,16]. Crystallization of β–SiC was observed above 1000 ºC, and at 1400 ºC C(graphite) and SiO_2(cristobalite) could be identified. The ceramic yield α attained 74 wt.% in the filler–free polysiloxane.

Filler–matrix reaction starts with dissolution of C in Ti to form a Ti–solid solution, followed by precipitation of a first generation of hyperstoichiometric titanium carbide, TiC_{1-x}(I), above 600 ºC with x = 0.48 at 600 and x = 0.33 at 1400 ºC, respectively. The maximum of the filler carburization rate strongly depends on the particle size of the filler powder and was found to lie at approximately 800 ºC for a mean particle size of 1–3 μm [7]. In contrast to the filler–free polymer, crystallization of the matrix glass in the Ti–filler loaded systems results in the formation of mainly SiO_2 (cristobalite) above 1000 ºC and only traces of SiC. Compared to the filler free system a significantly higher ceramic yield of the polymer phase of α = 80 wt% was found in the filler–loaded system.

Fig. 7 summarizes the microstructural transformations, as derived from XRD and TEM investigations, that occur during heating of a filler–free and a filler–loaded polysiloxane precursor in Ar–atmosphere. Compared to the filler–free polymer pyrolysis, the reactive filler mainly accounts for a significant reduction of shrinkage, a reduction of the C–content in the silicon oxycarbide matrix, a decrease of the devitrification temperature of the Si–O–C glass, and an increase of the SiO_2–fraction in the crystallized matrix.

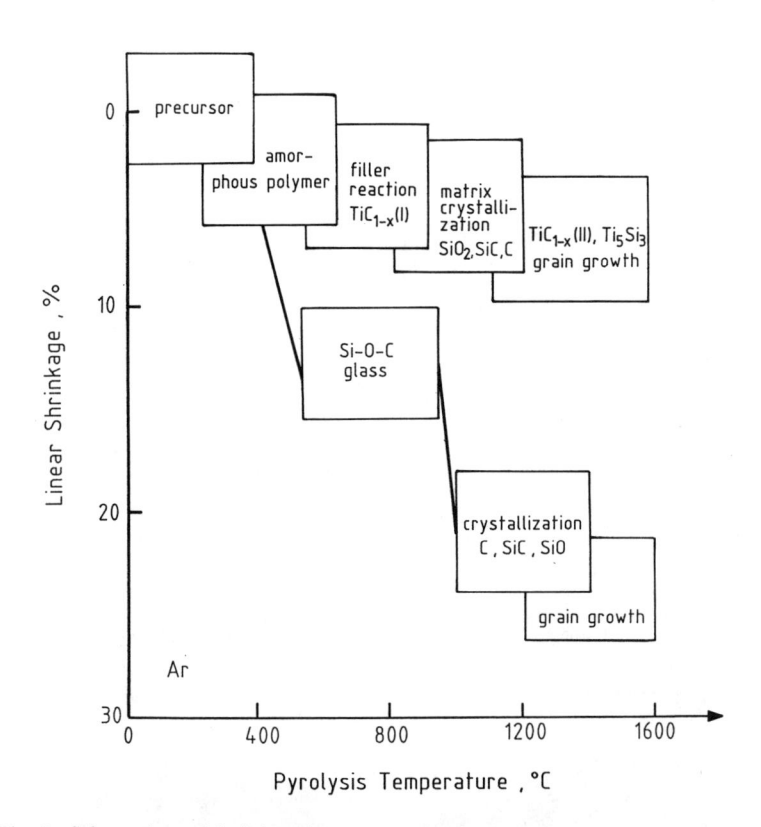

Fig. 7 Schematic model of the microstructural transformations during heating of a filler–free polysiloxane and a polysiloxane/titanium mixture in Ar–atmosphere.

Fig. 8 shows the mean crystallite sizes (determined from TEM micrographs) of the titanium carbide reaction products as a function of pyrolysis temperature. Below 800 °C titanium carburization seems to be nucleation–controlled whereas above this temperature grain growth results in a significant increase of particle size. Heat treatment at 1000 °C resulted in the formation of Ti_5Si_3 and a second generation of $TiC_{1-x}(II)$ at the filler matrix interface which may be explained by

$$3 SiC + 8 Ti \longrightarrow Ti_5Si_3 + 3 Ti_{1-x}(II) \qquad (15)$$

when unreacted Ti reacts with SiC precipitated from the matrix phase [17].

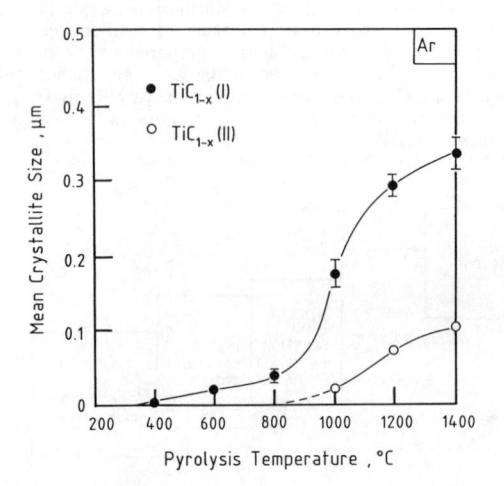

Fig. 8 Mean crystallite sizes of titanium carbide (I) and (II) as a function of the pyrolysis temperature.

Fig. 9a shows a TEM–BF micrograph of the ceramic composite obtained from a polysiloxane/20 vol.% titanium mixture pyrolyzed at 1400 ºC. The titanium carbide grains appear as well-faceted particles (dark) embedded in a matrix of silicon oxycarbide (bright). The polymer-derived silicon oxycarbide matrix exhibits a unique nanostructure with turbostratic carbon layers forming an interconnected network, Fig. 9b, which may provide a high electrical conductivity. While at temperatures above 1000 ºC microcrystalline titanium carbide inclusions are formed which often contain dislocations of the dominating gliding system $\{111\}<1\bar{1}0>$, nanocrystalline reaction layers are formed at lower temperatures at the filler/matrix interphase and along grain boundaries in the polycrystalline filler grains.

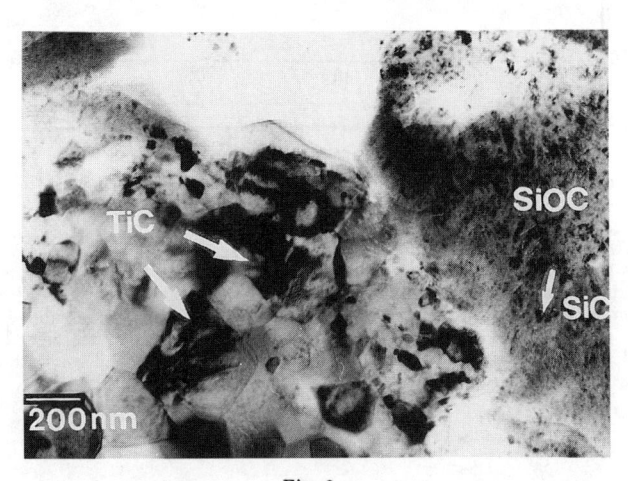

Fig. 9a

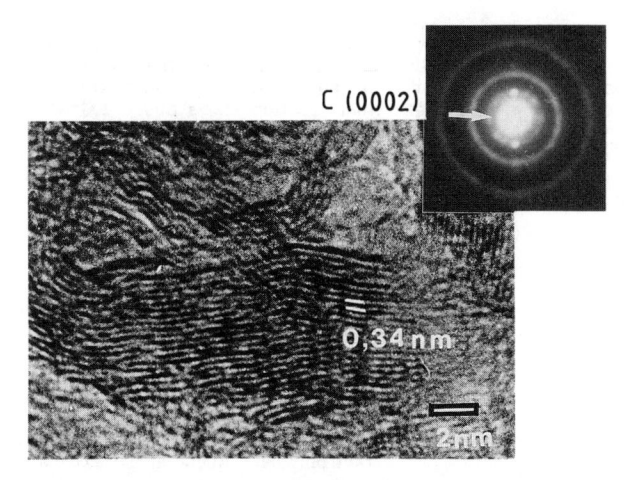

Fig. 9b

Fig. 9 TEM–BF–micrographs of the composite microstructure, a), and the polymer–derived matrix, b), after pyrolysis in Ar–atmosphere at 1400 °C.

Properties

Significant property changes can be expected as a result of the thermally induced microstructural transformations when percolation of the filler particles occurs, a continuous network of turbostratic carbon is formed and crystallization of the polymer–derived matrix starts. Thermal diffusivity or electrical conductivity [18], for example, will exhibit pronounced changes at the transformation temperatures which strongly depend on the polymer chemistry and the reactivity of the filler phase. The present study concentrates on the effect of heat treatment and filler loading on the mechanical behavior of the pyrolyzed materials.

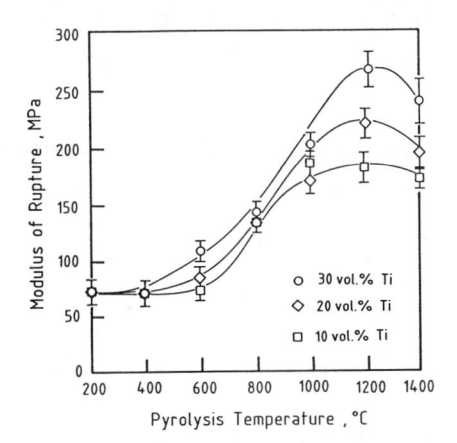

Fig. 10 Four–point bending strength of polysiloxane/titanium mixtures after pyrolysis at different temperatures in Ar–atmosphere.

Fig. 10 shows the development of strength (4–point bending 40/20 mm) as a function of pyrolysis temperature in the polysiloxane derived materials containing 10, 20 and 30 vol. % of the titanium powder in the initial mixtures. Strength is found to increase when polymer–filler reaction starts above 400 – 600 ºC and reaches a maximum at 1200 ºC when the reaction is completed. In addition to the strength the Weibull modulus also increases from m ≈ 9 for 10 vol.% filler up to m ≈ 15 for the specimen containing 30 vol.% of the filler powder, indicating a significant improvement of microstrucural homogeneity by the filler dispersion.

CONCLUSIONS

In general, the results of this study offer analytical and experimental support for the decisive role an active filler phase can play in affecting the dimensional change and porosity generation during formation of ceramic composites from polymer precursor systems. Although the physical and chemical polymer–filler interactions are not fully understood yet, the polymer decomposition chemistry is suggested to exert a strong influence on the structural relaxation behavior of the polymer–derived matrix which ultimately determines shape integrity of a bulk component. Hence, combination of different polymers, fillers, and reactive atmospheres may result in a variety of novel ceramic composite materials which can further be modified by additional inert fillers such as reinforcing particles or continuous fibers. Using well–established plastic forming techniques the active filler controlled polymer pyrolysis process (AFCOP) offers an attractive near–net shape fabrication route to complex shaped ceramic composite materials and components.

Rerences

[1] D. Seyferth, G.H. Wiseman, J.M. Schwark, Y.F. Yu, C.A. Poutasse, in Inorganic and Organometallic Polymers, ACS Symposium Series, vol. 360, 143 (1987)
[2] Y.F. Yu, T.I. Mah, in Better Ceramics Through Chemistry II, MRS Symposia Proceedings, vol. 73, 559 (1986)
[3] S. Yajima, J. Hayashi, M. Omori, K. Okamura, Nature 261, pp. 683 (1976)
[4] R.W. Rice, Am.Ceram.Soc.Bull. 62, 889 (1983)
[5] J. Lipowitz, J.A. Rabe, L.K. Frevel, R.L. Miller, J.Mat.Sci. 25 2118 (1990)
[6] M. Seibold, P. Greil, in Adv.Mat.Processing 1, edt. H.E. Exner, V. Schumacher, DGM Inform.Ges., Oberursel, FRG 641 (1990)
[7] P. Greil, M. Seibold, in Ceramic Transactions, Vol. 19, Advanced Composite Materials, edt. M.D. Sacks, The Am.Ceram.Soc. Westerville, OH, 43 (1991)
[8] D. Seyferth, N. Bryson, D.P. Workmann, C.A. Sobon, J.Am.Ceram.Soc. 74 2687 (1991)
[9] F.I. Hurwitz, P.J. Heimann, J.Z. Gyenkenyesi, J. Masnovi, X.Y. Bu, Ceram. Eng.Sci.Proc. 12 1292 (1991).
[10] T. Erny, M. Seibold, O. Jarchow, P. Greil, to be publ. in J.Am.Ceram.Soc. 75 (1992)
[11] M. Seibold, P. Greil, to be publ. in J.Europ.Ceram.Soc. 8 (1992)
[12] I. Barin, Thermochemical Data of Pure Substances, Verlag Chemie, Weinheim, Germany (1989)
[13] L.E. Toth, Transition Metal Carbides and Nitrides, Academic Press, London 1971
[14] P. Greil, M, Seibold, J.Mat.Sci. 27, 1053 (1991)
[15] G.M. Renlund, S. Prochazka, R.H. Doremus,J.Mat.Res. 6 2716 and 2723 (1991).
[16] G. Jenkins, K. Kawamura, Polymeric Carbons – Carbon Fiber, Glass and Char, Cambridge University Press (1976)
[17] M.B. Chamberlain, 72 305 (1980)
[18] M. Monthieux, A. Oberlin, E. Bouillon, Comp. Sci.Techn 37 21 (1990)

SYNTHESIS OF CARBON/FERRITE COMPOSITE BY IN-SITU PRESSURE PYROLYSIS OF ORGANOMETALLIC POLYMERS

SHIN-ICHI HIRANO, TOSHINOBU YOGO, KOICHI KIKUTA AND MAKOTO FUKUDA
Department of Applied Chemistry, Nagoya University,
Furo-cho, Chikusa-ku, Nagoya 464, Japan

ABSTRACT

Ferrite particle-dispersed carbon composite were synthesized by in-situ pressure pyrolysis of organometallic polymers at temperatures from 500 to 700°C under 125 MPa. Magnetite-dispersed carbon composite could be prepared from 550 to 700°C at 125 MPa. Nickel and nickel zinc ferrite particles were dispersed in carbon matrices at 550°C and 500°C, respectively, under 125 MPa. The morphology of the carbon matrix can be controlled by the pyrolysis conditions and the amount of coexistent supercritical water. Carbon spherulites of several micrometers dispersed with ferrite particles less than 100 nm were successfully synthesized by pressure pyrolysis of organometallic polymers in the presence of supercritical water. The saturation magnetization of magnetite-, nickel ferrite- and nickel zinc ferrite-dispersed carbon were 74, 30 and 65 emu/g, respectively. The coercive force of nickel ferrite-dispersed carbon was about 200 Oe.

INTRODUCTION

Organometallic polymers including metal-carbon bonds are versatile as starting materials for the synthesis of carbon/metal composites. Metal compound-dispersed carbon composites of controlled morphology have many applications such as shielding materials for electromagnetic waves, toners and catalysts. The pyrolysis of organometallic polymers for the synthesis of metal/carbon composites offers general advantages of high product purity and high dispersion of metallic compounds in the carbon matrices. In addition, the pressure pyrolysis has two remarkable features, 1) controllability of morphology for the carbon matrix and 2) high yield of carbon, which are distinct differences from the pyrolyses at ordinary or reduced pressure.

Hirano et al. [1,2] synthesized isotropic carbon spherulites in as high as 85% yield by pressure pyrolysis of divinylbenzene. A variety of metals, metal compounds and alloy particles can be dispersed in carbon matrices by the pressure pyrolysis of organometallic polymers [3-14]. The metal/carbon composites consist of carbon matrix of micrometer size and sub-micron-sized metal particles. The morphology of the carbon matrix was affected by the pyrolysis conditions of the parent polymers as well as the metal concentration in the starting polymers. The magnetic properties of metal-dispersed carbon reflect the crystallinity and the particle size of the metals, which have been found to depend strongly on the properties of both the carbon-metal bond of the organometallic compounds and the carbon-carbon bond of the polymer matrix.

This paper describes the processing and properties of carbon/ferrite composites prepared by pressure pyrolysis of organometallic polymers. Three kinds of ferrite particles, magnetite (Fe_3O_4), nickel ferrite ($NiFe_2O_4$) and nickel zinc ferrite ((Ni,Zn)Fe$_2$O$_4$), could be dispersed in the carbon matrices. Spherulitic carbon composites dispersed with ferrite particles can be synthesized by the pressure pyrolysis of organometallic polymer and supercritical water mixtures.

EXPERIMENTAL PROCEDURE

Starting Materials

Vinylferrocene (VF), nickelocene (Nc), and zinc acetylacetonate (ZA) were used as starting organometallic compounds. Commercially available divinylbenzene (a mixture of 55% m- and p-divinylbenzene and 45% ethylbenzene) was employed for the synthesis of the starting polymer matrix. The molecular structures are shown in Fig. 1.

Synthesis of Ferrite-dispersed Carbon

Pressure pyrolyses were carried out in a hydrothermal apparatus of the cold-seal type. Figure 1 also illustrates the scheme for processing the carbon composites. Organometallic compounds corresponding to the composition of ferrites (NiFe$_2$O$_4$, Ni$_{0.5}$Zn$_{0.5}$Fe$_2$O$_4$) were dissolved in divinylbenzene (DVB) under nitrogen. The solution was then sealed into a thin-walled gold capsule of 3.0 or 5.0 mm in diameter and 50 mm in length.

Various solutions of DVB-containing organometallic compounds were polymerized at 300°C and 100 MPa for 2h and then pyrolyzed at temperatures from 400 to 700°C and 125 MPa for 3h. A weighed amount of water was sealed with the solutions or polymers before pyrolysis in case of the synthesis of ferrite-dispersed carbon. The temperature was raised at 10°C/min at a constant pressure of 125 MPa. The pressure was kept isobaric by releasing water as a pressure transporting medium during heating. Specimens were quenched after each experimental run.

Characterization of Metal-dispersed Carbons

The yield of carbon was as high as about 75 wt% of the starting copolymer. The carbonized product was analyzed by X-ray diffraction (XRD) using CuKα radiation with a monochromater. The lattice constant of ferrite was determined from its (440) and (511) X-ray diffractions with tungsten powder as an internal standard. The synthesized metal-dispersed carbons were characterized by scanning electron microscopy (SEM) and transmission electron microscopy (TEM).

The thermomagnetic measurement was conducted on ferrite-dispersed carbon composites using a magnetic balance from room temperature to 600°C. The saturation magnetization and the coercive force of the ferrite-dispersed carbon composites were measured by a vibrating sample magnetometer (VSM) at room temperature.

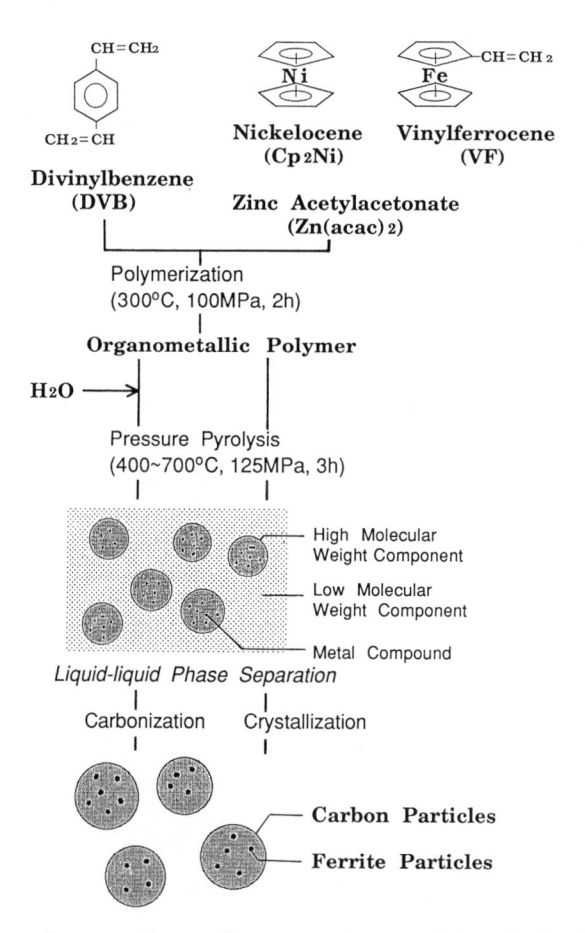

Fig. 1 Scheme for preparing carbon composite containing finely dispersed ferrites.

RESULTS AND DISCUSSION

Synthesis of Organometallic Polymers

The organometallic compounds should be soluble in monomer compounds for molecular-level mixing. Vinylferrocene (VF) shown in Fig. 1 was copolymerized with DVB to form organoiron copolymers. The DVB-VF copolymer has a cross-linkage via benzene rings, since DVB has two vinyl groups on a benzene ring as shown in Fig. 1. On the other hand, nickelocene and zinc acetylacetonate has no vinyl group (Fig. 1) for the copolymerization with DVB. However, an unsaturated monomer solution of DVB containing the organometallic compounds is polymerized under pressure yielding polymers dispersed with organometallic compounds at the molecular level.

Synthesis of Ferrite-Dispersed Carbon

The XRD profiles of the products formed from DVB-5.1mol%VF at temperatures from 550 to 700°C and 125 MPa are shown in Fig. 2. DVB-VF copolymer was pyrolyzed at 550°C and 125 MPa producing cementite (Fe_3C)-dispersed carbon as shown in Fig. 2a [3]. The dispersion of cementite was also confirmed by its Curie temperature in the thermomagnetic measurement. On the other hand, the pressure pyrolysis of the organoiron copolymer in the presence of water at 125 MPa above 550°C yields carbon dispersed with magnetite (Fe_3O_4) [8]. The broad X-ray diffraction centered at $2\theta=25.5$ degrees is derived from turbostratic carbon. The sharp reflections are attributed to those of magnetite. No cementite was formed in the product synthesized even at 700°C (Fig. 2c). The Curie temperature of magnetite-dispersed carbon was 585°C, which was in good agreement with the reported value [8].

Figures 3 and 4 show the XRD profiles of the product from DVB-6.0mol%VF-3.0mol%Nc and DVB-6.0mol%VF-1.5mol%Nc-1.5mol%ZA, respectively, pyrolyzed between 500 and 700°C at 125 MPa. No highly crystalline phase was formed at 550°C in the absence of water as shown in Fig. 3a. Nickel ferrite was formed in a carbon matrix as the single phase at 550°C by the addition of water (Fig. 3b). However, iron-nickel carbide appeared at $2\theta=44$ degrees by pyrolysis at 600°C and increased in amount with increasing pyrolysis temperature to 700°C (Fig. 3d). The lattice constant of nickel ferrite formed at 600°C was 837 pm, which corresponds to the composition between $NiFe_2O_4$ ($a_0=833.9$ pm) and Fe_3O_4 ($a_0=839.6$ pm). The composition of the Ni ferrite was calculated to be $Ni_{0.61}Fe_{0.39}O_4$ based upon the Vegard's rule. Therefore, Ni-Fe carbide is considered to have nickel rich composition.

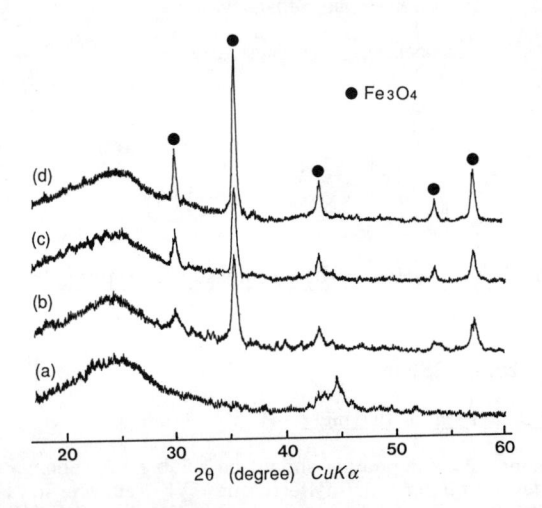

Fig. 2 XRD profiles of magnetite-dispersed carbons formed from DVB-5.1mol%VF at 125 MPa.
(a) 550°C, no water, in the presence of 15wt% water at (b) 550°C, (c) 600°C, (d) 700°C.

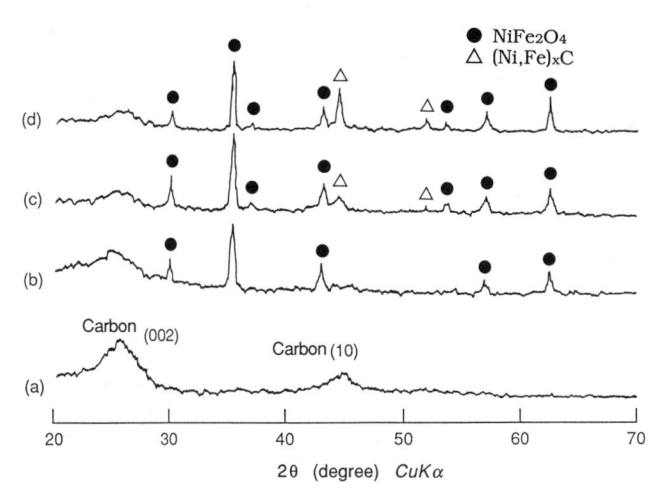

Fig. 3 XRD profiles of nickel ferrite-dispersed carbons formed from DVB-6.0mol%VF-3.0mol%Nc at 125 MPa.
(a) 550°C, no water, in the presence of 15wt%H$_2$O at (b) 550°C, (c) 600°C, (d) 700°C.

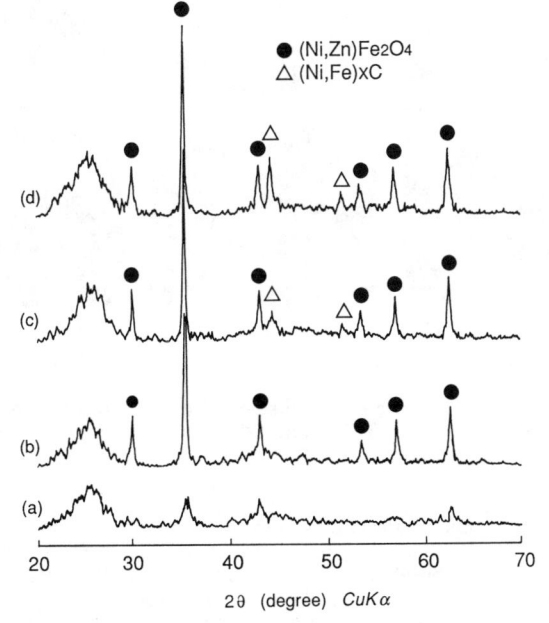

Fig. 4 XRD profiles of nickel zinc ferrite-dispersed carbons synthesized from DVB-6.0mol%VF-1.5mol%Nc-1.5mol%ZA.
(a) 500°C, no water, in the presence of 30wt%H$_2$O at (b) 500°C, (c) 600°C, (d) 700°C

Similarly, the single phase of nickel zinc ferrite was formed at 500°C, and a small amount of iron-nickel carbide was observed in the pyrolysis product above 600°C as shown in Fig. 4 [13]. The lattice constant of nickel zinc ferrite was 840.8 pm, which corresponds to a Zn rich composition compared to the starting composition of $Ni_{0.5}Zn_{0.5}Fe_2O_4$.

Morphology of Carbon Matrix

The representative morphology of metal-dispersed carbon is coalesced polyhedra of irregular shape [3]. When the concentration of water is low, carbon dispersed with ferrite has also the polyhedral structure. However, the morphology of magnetite-dispersed carbon changes from coalescing polyhedra to spherulites with increasing water concentration from 0 to 20 wt% at the same pyrolysis temperature [8]. A similar change from lump structure to spherulitic carbon was observed for nickel zinc ferrite-dispersed carbon as the pyrolysis temperature increased from 450 to 700°C at a constant water concentration of 30 wt% (Fig. 5) [13]. The diameter of carbon spherulites synthesized at 700°C ranged from 2 to 5 μm.

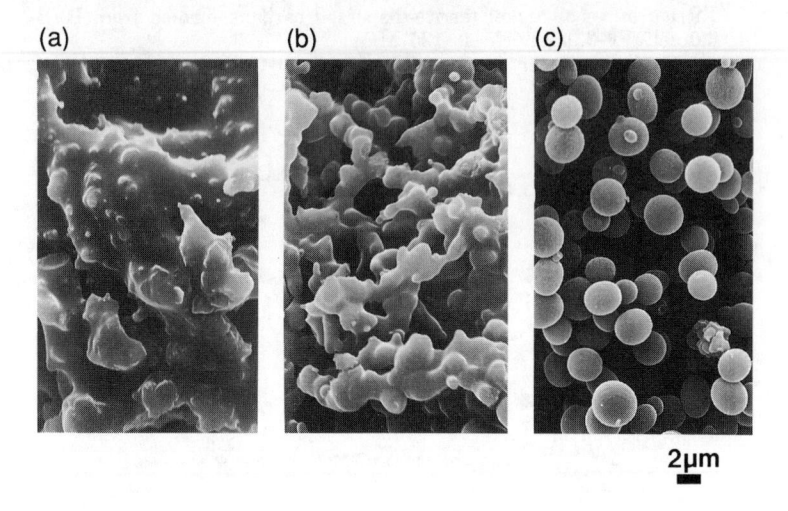

Fig. 5 SEM of nickel zinc ferrite-dispersed carbon formed from DVB-6.0mol%VF-1.5mol%Nc-1.5mol%ZA at 125 MPa in the presence of 30 wt% water. (a) 450°C, (b) 500°C, (c) 700°C.

Spherulites and coalescing spherulites are formed by pressure pyrolysis of poly-DVB itself, depending upon the pyrolysis pressure and temperature [1,2]. The thermal breakdown of cross-linked poly-DVB affords low molecular weight oligomers at the initial stage of pyrolysis. The oligomers of different molecular weight undergo the liquid-liquid microphase separation of immiscible phases yielding spherulitic carbon of round shape. However, metal particles formed by aggregation during pyrolysis change the phase separation behavior from homogeneous to heterogeneous.

The morphology of the carbon was found to be controlled by the concentration of water and the pyrolysis temperature. Spherulitic carbon is successfully synthesized by the proper selection of the water content and temperature. Supercritical water is also considered to act as a low molecular weight component giving a homogeneous microphase separation system.

Figure 6 shows the microstructure of nickel ferrite-dispersed carbon formed from DVB-6.0mol%VF-3.0mol%Nc-30wt%H$_2$O at 550°C and 125 MPa. The size of ferrite particles was below 30 nm.

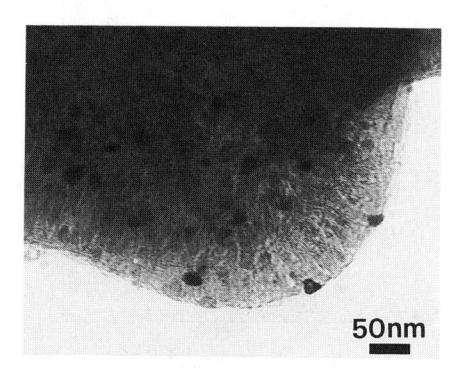

Fig. 6 TEM photograph of nickel ferrite particles dispersed in a carbon matrix prepared from DVB-6.0mol%VF-3.0mol%VF-15wt%H$_2$O at 550°C and 125MPa.

Magnetic Properties of Ferrite-Dispersed Carbon

Figure 7 summarizes the saturation magnetization (σ_s) of ferrite-dispersed carbon synthesized from organometallic polymers at temperatures between 400°C and 700°C at 125 MPa. The σ_s of the specimen was converted to the value per gram of ferrites. The σ_s of nickel ferrite-dispersed carbons increases with increasing pyrolysis temperature, and reached a constant value of 57 emu/g above 600°C. Nickel zinc ferrite-dispersed carbon shows constant σ_s above 500°C. Magnetite-dispersed carbon also shows a constant value above 550°C. When the pyrolysis temperature was the same, the σ_s of magnetite-dispersed carbon was slightly higher than that of nickel zinc ferrite-dispersed carbon. The σ_s of nickel ferrite-dispersed carbon was much smaller than those of magnetite- and nickel zinc ferrite-dispersed carbon.

The reported values of σ_s for ferrites increase in the order of NiFe$_2$O$_4$ (50 emu/g) < Fe$_3$O$_4$ (92) < Ni$_{0.5}$Zn$_{0.5}$Fe$_2$O$_4$ (108). The σ_s of magnetite-dispersed carbon shows about 80 % of the reported value. The σ_s of nickel ferrite- and nickel zinc ferrite-dispersed carbon formed at 550 and 500°C shows about 60 and 59 % of the reported value, respectively. Iron, nickel and zinc cations are not considered to distribute properly at the tetrahedral and octahedral site of ideal spinel structure in nickel ferrite and zinc ferrite particles in carbon matrix. Nickel iron carbide contributes to the σ_s of carbons dispersed with nickel ferrite and nickel zinc ferrite formed above 600°C. When metal carbide is formed, the composition of nickel ferrite and nickel zinc ferrite deviates from the starting composition as described in the preceding section. Since the σ_s of nickel zinc ferrite shows the maximum value at $Ni_{0.5}Zn_{0.5}Fe_2O_4$ composition, the deviation of composition from Zn:Ni=0.5:0.5 to zinc rich is responsible for the lower σ_s than that of magnetite.

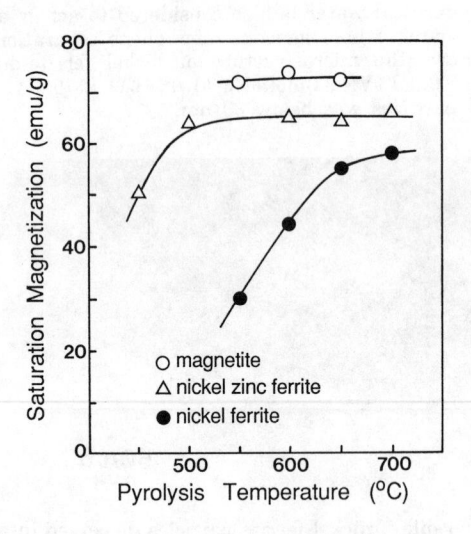

Fig. 7 Saturation magnetization of ferrite-dispersed carbon formed at various temperatures and 125 MPa.
○: magnetite, ●:nickel ferrite, △:nickel zinc ferrite.

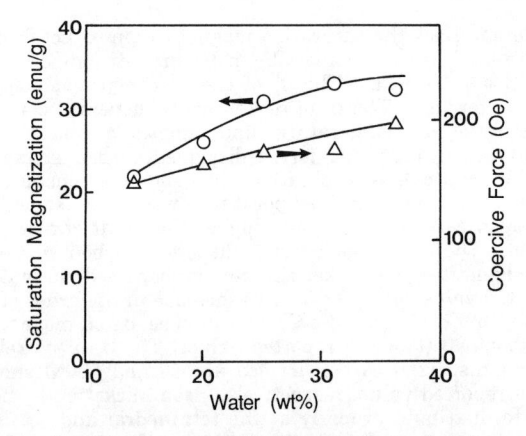

Fig. 8 Change of saturation magnetization and coercive force of nickel ferrite-dispersed carbon with concentration of water.

Figure 8 shows the relation between the amount of water and the σ_s of nickel ferrite dispersed-carbon formed from DVB-6.0mol%VF-3.0mol%VF-15wt%H$_2$O at 550°C and 125 MPa. The σ_s increases with increasing water content from 13 to 35 wt%. The coercive force (Hc) of nickel ferrite-dispersed carbon is also plotted as a function of the water content in Fig. 8. The maximum coercive force of nickel ferrite-dispersed carbon is 200 Oe at 36 wt% water. Water promotes the crystallization of ferrite, increasing the crystallite size and regularity, and leading to the increases in both σ_s and Hc.

CONCLUSIONS

Organometallic polymers were pressure pyrolyzed to synthesize carbon composites with finely dispersed ferrite particles in high yields. The size and the morphology of the carbon matrix can be controlled by the selection of the pyrolysis conditions as well as the amount of coexistent water and the metal concentration in the polymers. Spherulitic carbons dispersed with ferrite particles were synthesized by the selection of water content and pyrolysis temperature. The single phase particles of nickel ferrite or nickel zinc ferrite could be dispersed in carbon at 500°C, while carbon dispersed with single phase magnetite were obtained from 550 to 700°C. The saturation magnetization and coercive force of ferrite-dispersed carbon were dependent upon the pyrolysis conditions of the parent organometallic polymers as well as the water concentration. This processing affords a novel method for producing carbon/ferrite nano-composites with controlled microstructure and morphology.

REFERENCES

[1] S.Hirano, F.Dachille and P.L.Walker Jr, *High Temp. High Press.* **5** 207 (1973).
[2] S.Hirano, M.Ozawa and S.Naka, *J. Mater. Sci.* **16** 1989 (1981).
[3] S.Hirano, T.Yogo, H.Suzuki and S.Naka, *ibid.* **18** 2811 (1983).
[4] T.Yogo, S.Naka and S.Hirano, *ibid.* **24** 2071 (1989).
[5] S.Hirano, T.Yogo, N.Nogami and S.Naka, *ibid.* **21** 225 (1986).
[6] T.Yogo, E.Tamura, S.Naka and S.Hirano, *ibid.* **21** 941 (1986).
[7] T.Yogo, H.Yokoyama, S.Naka and S.Hirano, *ibid.* **21** 2571 (1986).
[8] T.Yogo, S.Naka and S.Hirano, *ibid.* **24** 2115 (1989).
[9] T.Yogo, S.Naka and S.Hirano, *ibid.* **22** 985 (1987).
[10] S.Hirano, T.Yogo, K.Kikuta and S.Naka, *ibid.* **21** 1951 (1986).
[11] T.Yogo, H.Tanaka, S.Naka and S.Hirano, *ibid.* **25** 1719 (1990).
[12] T.Yogo, H.Suzuki, H.Iwahara, S.Naka and S.Hirano, *ibid.* **26** 1363 (1991).
[13] S.Hirano, T.Yogo, K.Kikuta and M.Fukuda, *Ceram. Trans.* **22** 33 (1991).
[14] S.Hirano, T.Yogo, K.Kikuta, M.Takase and M.Fukuda, *Jpn. J. Chem. Soc.,* **1991** 1261 (1991).

MAGNETIC PROPERTIES OF MECHANICALLY ALLOYED NANO-CRYSTALLINE Cu/Fe COMPOSITES

C.P. REED[*], S.C. AXTELL[*], R.J. DE ANGELIS[*],
B.W. ROBERTSON[*], V.V. MUNTEANU[*] AND S.H. LIOU[**]
Center for Materials Research and Analysis and [*]Dept. of Mechanical Engineering,
[**] Dept. of Physics and Astronomy, University of Nebraska-Lincoln, Lincoln, Nebraska 68588-0656

Abstract

Metal powders of the composition 70 at% Cu and 30 at% Fe were produced by high energy mechanical alloying of the elemental powders. The powders were processed in a Spex 8000 mixer/mill for various times to investigate the potential of the mechanical alloying process for producing nano-composite structures with modified magnetic properties. Optical microscopy revealed a layered structure of alternating copper and iron that developed upon milling. The spacing between the layers decreased with milling time, becoming optically unresolvable (< 1 μm) after four hours of milling. A single profile x-ray diffraction profile shape analysis technique was used to determine the average diffracting particle size of the copper and iron phases. The diffracting particle size decreases with alloying time reaching values of 7.5 nm and 2 nm, for copper and iron respectively, after eight hours of alloying. The magnetic coercivity increased with milling time initially, reaching a maximum value above 300 Oe after six hours of milling. These results are discussed and compared to results obtained in Ag/Fe and Cu/Fe nano-composite films.

Introduction

Metal/metal composite magnetic materials have been developed to take advantage of the magnetic properties of fine metal particle dispersions. The magnetic properties of the composites can be manipulated by controlling the size and dispersion of the magnetic particles. In this investigation, iron and copper powders were mechanically alloyed to produce nano-composite powders. The structure of the powders, characterized employing optical microscopy and x-ray diffraction, was correlated with the magnetic properties determined from the same samples.

The Fe/Cu powder mixes were selected for this investigation because iron is magnetic and the phase diagram indicated that these elements have very limited solid solubility. The maximum solubility of copper in iron is about 12 at.% at around 1400°C. As the temperature is decreased the solubility rapidly decreases to basically zero at the ambient temperature of milling. For this reason, it is expected that the phases will remain separate during processing of the composite powders and a fine dispersion of the magnetic iron particles in the nonmagnetic copper matrix will be formed.

Experimental Procedure

Composite metal powders composed of iron and copper were produced by mechanical alloying of the elemental powders. The powders were alloyed under air utilizing a Spex 8000 mixer/mill for times up to 12 hours. The starting iron powder, Federal Mogul ARMCO iron, was spherical in shape with a sieve range of -100 to +200 mesh. The copper powder, an electrolytic copper purchased from Fisher Scientific, was in the form of platelets 30-80 μm by one or two microns in thickness. Scanning electron micrograph images of the starting powders are shown in Fig. 1. A series of Cu_{70}-Fe_{30} powders were produced by alloying with three tungsten carbide grinding balls in a tungsten carbide lined Spex grinding chamber containing a powder charge of 13 gms. giving a charge ratio of 2.5.

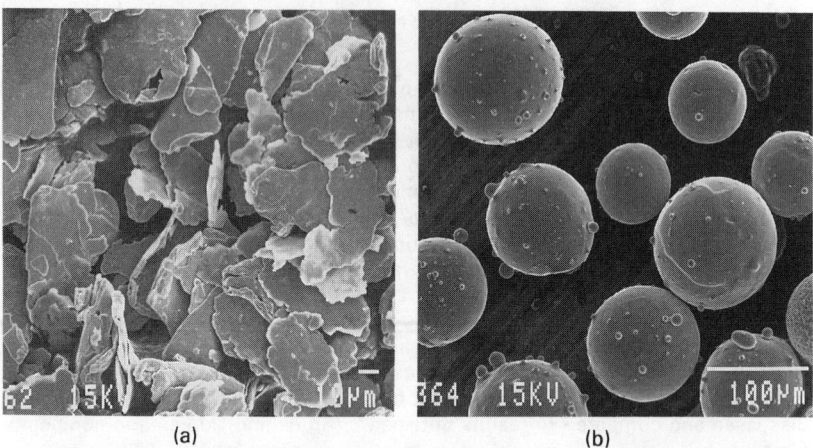

Fig. 1. Scanning Electron Micrographs of the Starting Powders: (a) Copper and (b) Iron.

The structural characteristics of the milled powders were determined by optical metallography and x-ray diffraction. The x-ray patterns were obtained on the Philips diffractometer employing a copper tube and outfitted with a diffracted beam graphite monochromator. The x-ray scans were collected using a 0.2° receiving slit and a 1°-4° variable divergent slit. Magnetic properties of the milled powders were determined with a vibrating sample magnetometer and a MicroMag Model 2900 alternating gradient force magnetometer.

Experimental Results and Discussion

As the copper and iron powders are milled they tend to form layered structures. The average spacing of the layers in the structure becomes smaller as the time of milling increases. Optical micrographs showing the types of layered structure formed during milling are shown in Fig. 2. The layered structure, very evident at short milling times, develops a fine spacing (<1 micron) and becomes unresolvable optically at longer milling times. The layers are believed to break up, forming high aspect ratio separated particles.

X-ray diffraction patterns obtained from the mechanically milled Fe/Cu powders are shown in Fig. 3. A quick examination of these patterns shows the milled iron is in the body centered cubic phase and copper is face centered cubic. The widths of the diffraction peaks increase with milling time, however the broadening behavior of the iron and copper phases are quite different. The average

crystallite size was determined from the broadened x-ray line profiles using a
Fourier deconvolution method which has been reported previously [1,2].

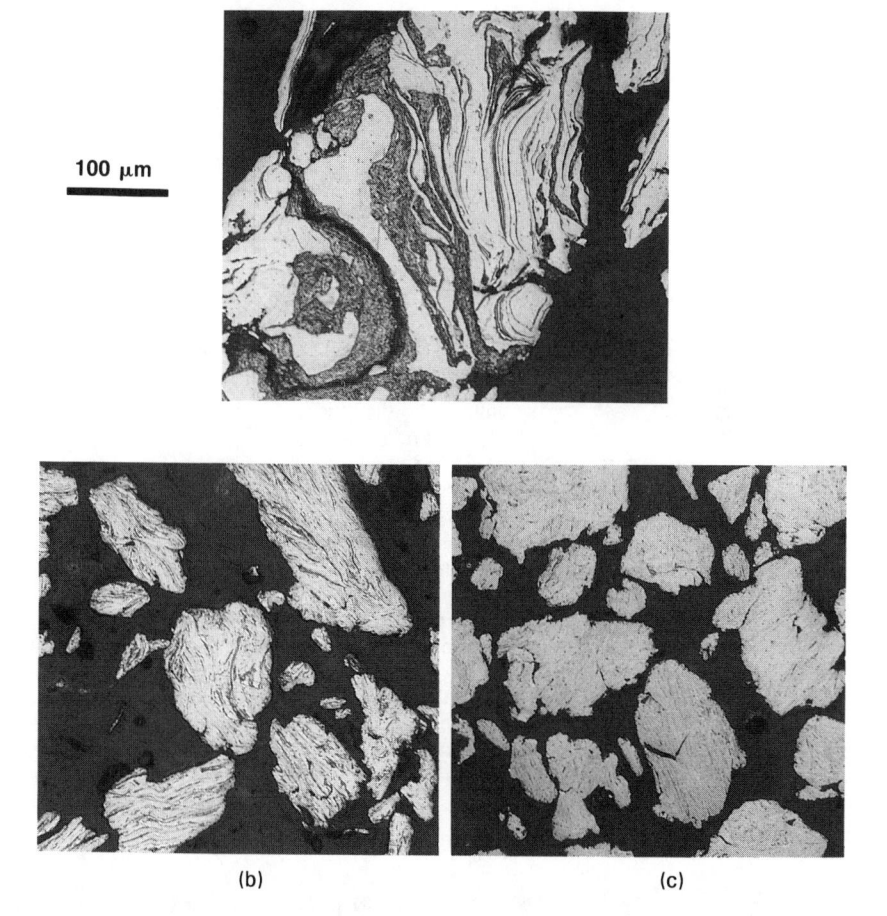

Fig. 2. Optical Micrographs Showing the Refinement of the Layer Structure During Mechanical Alloying of Cu_{70}-Fe_{30}. Milling Times Were; (a) One Hour, (b) Three Hours, and (c) Six Hours.

The average diffracting crystallite sizes of the Fe/Cu powders as a function of milling time are shown in Fig. 4. The iron crystallite size decreases with increasing milling time from over 18 nm to 2.1 nm after eight hours of milling. The copper crystallite size decreases rapidly with milling time from about 13 nm to 7.5 nm in the initial five hours of milling. Longer milling times, up to 12 hrs., had no observable affect on the diffracting particle size.

The decrease in diffracting crystallite size in both of the metallic phases is due to the action of the impacting balls with each other and the chamber walls. These collisions which plastically deform the powders produce an increase in the defect density of the material.

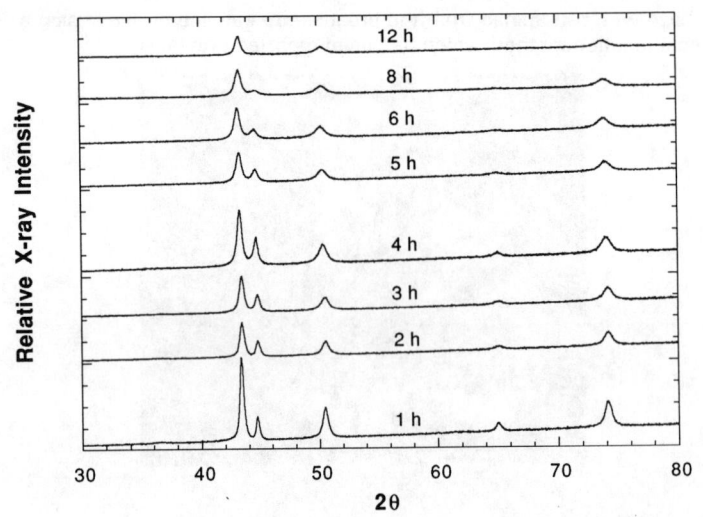

Fig. 3. X-Ray Diffraction Patterns of Mechanical Alloyed Cu_{70}-Fe_{30} at the Milling Times Indicated on Each Pattern.

The increase in defect density reduces the average length of coherent diffracting crystal or diffracting particle size. Due to the fcc structure of copper, the internal energy of the defect density stored in the structure is less then the defect density possible to store in bcc iron. This difference in ability to store defects is reflected in the much shorter milling times to reach defect density saturation in copper than in iron.

The structure of the copper particle appears to saturate in defects when the value of the diffracting particle size is about 7.5 nm. As the milling time increases, the powders go through a phase of welding dominance in which the powder particles grow in size [3]. Further extension of milling times lead to such high levels

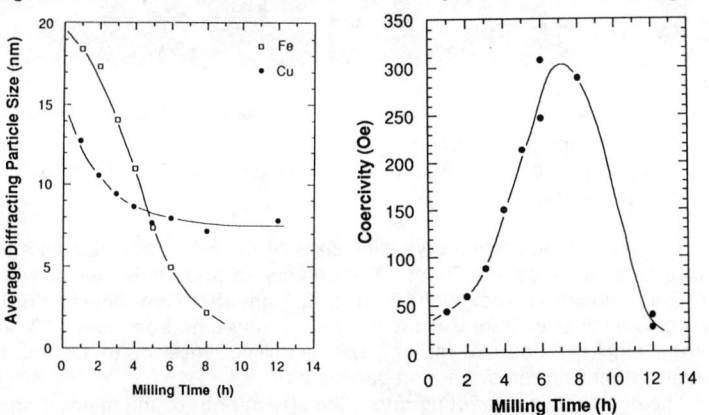

Fig. 4. Average Diffracting Particle Size of Copper and Iron Vs: Milling Time.

Fig. 5. Magnetic Coercivity of $Fe_{30}Cu_{70}$ as a Function of Milling Time.

of strain hardening that particle fragmentation becomes dominate resulting in a considerable reduction in the physical size of the particle.

As the milling time increases, the coercivity of the powders increases to a maximum as can be seen in the data shown in Fig. 5 The milled Fe/Cu powders demonstrates a maximum coercivity of over 300 Oe after six hours of milling. This maximum in coercivity occurs when the average diffracting crystallite size of the iron is 10 nm, very near to the single domain size for iron as reported by Luborsky [4].

Dislocation densities can be calculated from the average crystallite size, D_e^2, by assuming that crystallites are separated by single dislocations, then the density of dislocations is given by $1/D_e^2$. Using this relationship dislocation densities were calculated for the various milling times and are shown in Fig. 6. The dislocation densities for iron increase with milling time from 2.9×10^{11} cm^{-2} at one hour to 2.1×10^{13} cm^{-2} at eight hours. The copper dislocation densities were determined to be 6.2×10^{11} cm^{-2} after one hour of milling and remain at 2×10^{12} cm^{-2} after five hours of milling (see Fig.6).

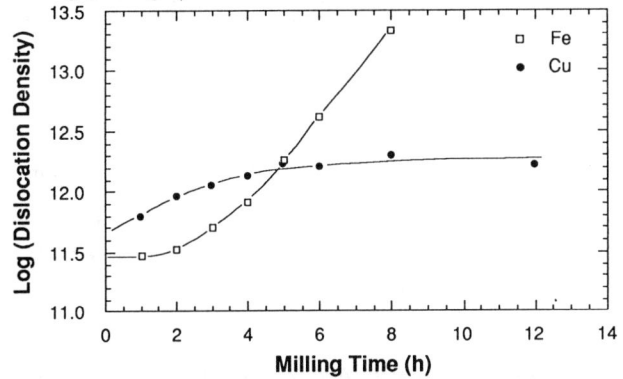

Fig. 6. Dislocation Densities of Cu_{70}-Fe_{30} as a Function of Milling Time.

The characteristic layered structure of alternating iron and copper becomes finer as alloying time increases, decreasing the size of the individual phases. The copper phase reaches an value of work hardening as shown by the crystallite size and dislocation density data, which remain constant after five hours of alloying. The iron phase, however, is still work hardening after milling eight hours. The data set terminates at eight hours because at longer alloying times the crystallite size becomes to small to produce resolvable diffraction peaks. The dislocation density determined from the diffracting particle size of the iron after eight hours of alloying was very high, 2.3×10^{13} cm^{-2}. In fact, it becomes difficult to define the iron crystallinity since the average crystallite size is less than 2.2 nm, a factor of five smaller than the single magnetic domain size. The coercivity of the Fe/Cu powders following long alloying times is adversely affected by the extensive damage inflicted on the crystallinity. An average crystallite size of 2.2 nm is associated with a size distribution function which would be expected to contain a significant number of iron particles much smaller than the single domain size. These particles would display superparamagnetic behavior causing the powders milled for extended times to demonstrate the observed decrease in coercivity. Even for alloying times of three to four hours where the coercivity is at a maximum value of about 300 Oe there may be a strong affect due to the existence of small particles formed by fragmentation. This may account for differences between the data collected in this investigation and the 900 Oe coercivity observed in Fe/Ag composite thin films [5,6] or the 500 Oe coercivity seen in Fe/Cu nano-composite films [7]. In these latter cases the maximum values also occurred close to 10 nm diffracting particle size; however the size distribution functions would be expected to be sharper than those developed by mechanical alloying. Therefore there would be a lesser fraction of particles with drastically reduced dimensions.

Conclusions

The results obtained in this investigation allow the following conclusions to be made concerning the usefulness of mechanical alloying as a method to process nano-composites of iron and copper for the modification of the magnetic properties.

* The average crystallite size of the iron and copper phases decreases with increased alloying time.

* The Cu_{70}-Fe_{30} powders showed a systematic variation in magnetic coercivity with iron crystallite size.

* The dislocation density in the powders increases with alloying time.

* The maximum coercivity in the powders is lower than that seen in similar Fe/metal nano-composites produced in thin film form by sputtering.

References

1. B.E. Warren, X-Ray Diffraction, Addison-Wesley, New York, 1969.

2. H.K. Kuo, P. Ganesan and R.J. De Angelis, A Method to Study Sintering of a Supported Metal Catalyst, Microstructural Science, vol. 8, Elsevier North Holland, 1980.

3. J.S. Benjamin and T.E. Volin, The Mechanism of Mechanical Alloying, Metall. Trans., vol. 5, 1974, p. 1929.

4. F.E. Luborsky, High Coercive Materials Development of Elongated Particle Magnets, J. Appl. Phys., vol 32, no. 3, 1961, p. 171S.

5. C.P. Reed, R.J. De Angelis, Y.X. Zhang and S.H. Liou, Substructure-Magnetic Property Correlation in Fe/Ag Composite Thin Films, Advances in X-Ray Analysis, Vol. 34, Plenum Press, 1991, p. 557.

6. Y.X. Zhang, S.H. Liou, R.J. De Angelis, K.W. Lee, C.P. Reed and A. Nazareth, The Process Controlled Magnetic Properties in Nanostructured Fe/Ag Composite Films, J. Appl. Phys., vol. 69 (8), 1991, p. 5273.

7. J.R. Childress, C.L. Chien and M. Nathan, Granular Iron in a Metallic Matrix, Appl. Phys. Lett., vol. 56 (1), 1990, p. 95.

Author Index

Archer, A.C., 17
Assink, Roger A., 85
Axtell, S.C., 177

Baney, Ronald H., 115
Bargon, Joachim, 47
Bauer, Barry J., 59
Baumann, Reinhard, 47
Beaucage, Greg, 85
Bergstrom, D.F., 31
Briber, Robert M., 59
Burns, G.T., 31

Cohen, Claude, 59
Corain, B., 103
Corvaja, C., 103
Cotterell, Brian, 3
Crowson, Andrew, 53

De Angelis, R.J., 177
Decker, G.T., 31
DeGroot, D.C., 133
Durall, R.L., 31

Ellsworth, Mark W., 67
Erny, Tobias, 155

Fryrear, D., 31
Fujii, Takashi, 141
Fukuda, Makoto, 167

Gentle, Theresa E., 115
Gornowicz, G.A., 31
Greil, Peter, 155
Guo, Fenchun, 11

Han, Xiaozu, 11
Hatano, Hiraku, 141
Hedrick, J., 37
Hirano, Shin-Ichi, 141, 167
Hofer, D., 37

Jerabek, K., 103
Jones, Phillip L., 53

Kanatzidis, M.G., 133
Kannewurf, C.R., 133
Katz, J.D., 149
Kikuta, Koichi, 167

Labadie, J., 37
Liou, S.H., 177
Liu, Y.-J., 133

Lora, S., 103
Lovell, P.A., 17

Mai, Yiu-Wing, 3
Malone, Shawn, 59
Mark, James E., 77
McDonald, J., 17
McKittrick, J., 149
Munteanu, V.V., 177
Muragaki, Hironobu, 141

Naslund, Robert A., 53
Nellis, W., 149
Novak, Bruce M., 67

Odagiri, N., 31
Olivier, Bernard J., 85

Palma, G., 103
Pecora, R., 109

Qui, Liying, 25

Reed, C.P., 177
Ren, Zhongyuan, 25
Robertson, B.W., 177
Russell, T., 37

Saam, John C., 91
Schaefer, Dale W., 85
Schindler, J.L., 133
Schmidt, Helmut K., 121
Seibold, Michael, 155
Sherratt, M.N., 17

Tokunoh, M., 31
Tracy, Mark A., 109
Tunaboylu, B., 149

Ulibarri, Tamara A., 85

Wakharkar, V., 37
Wang, Shuhong, 77
Wen, Jianye, 77
Wu, C.-G., 133
Wu, Jingshen, 3

Xu, Ping, 77

Yogo, Toshinobu, 167
Young, R.J., 17
Yun, Zhankui, 11

Zecca, M., 103

Subject Index

ablation, excimer laser, 47
alloying, mechanical, 177
alumina, 121, 149
    lanthanum beta, 141

biaxial extension, 77
blend, 3, 17, 25, 53, 59
block copolymer, 37
butadiene-acrylonitrile copolymer, 11

capacitor, 25
carbon, 167
cavitation, 3
Ce-TZP, 141
coating, 121
colloids, gold, 121
composite, 31, 85, 103, 121
    carbon, 167
    ceramic, 155
    inorganic-organic, 121
    liquid, 109
conductivity, 133
copolymer, 11, 37
    block, 37
    butadiene-acrylonitrile, 11
copper, 177
corrosion, 115
crosslink, 59
crosslinked, 103

deformation, 17
diffraction grating, 121
diffusion, 109

elastomer, silicone, 91
electrically conducting polymers, 47
electroplating, 47
elongation, 77
emulsion, 17
encapsulation, 25
epoxy, 25
excimer laser ablation, 47
extension, biaxial, 77

ferrite, 167
filled, 85
filler, 77, 91
    active, 155
foam, 37
fracture, 17
    static, 3
    strength, 141
    surface, 3
    toughness, 3, 141

gold, 121

high temperature, 37
HTBN, 11

hydrothermal stability, 141
hydroxy-terminated, 11

impact, 3
    modifier, 3
impedance spectroscopy, 133
in situ
    filled material, 85
    formation of filler, 91
    polymerization, 133
    precipitation, 77
interfacial energy, 91
interpenetrating polymer network, 59, 67, 103
iron, 177

J-integral, 3

lanthanum beta alumina, 141
latex, 17
ligand, bifunctional, 121
liquid crystalline, 53
lithographic patterns, 47

magnetic properties, 177
magnetization, saturation, 167
mechanical properties, 77
metal powder, 177
methacrylic acid, 121
microscopy, 17
    scanning electron, 3, 11
    transmission electron, 3, 11, 37
microwave sintering, 149
milling, 177
miscible, 59
modifier, 31
modulus, 77

nanocomposite, 115, 121, 177
    silica/silicone, 115
nanocrystalline, 149
nanopore, 37
network, 59, 67, 77, 103
non-shrinking, 67

ORMOCER, 121
organometallic polymer, 103
oxidative polymerization, 47

PALS, 53
particle, 17
    growth, 121
phase separation, 31
photolithography, 121
plasma sintering, 149
plastic flow, 3
polybutadiene, 25
polybutylene terephthalate, 3
polycarbonate, 3, 47
polycarbosilane, 155

poly(chloroacrylonitrile), 47
polydimethylacrylamide, 103
polydimethylsiloxane, 77, 85, 91, 115
polyetherimide, 53
polymer(s)
  bound, 167
  electrically conducting, 47
  organometallic, 167
  preceramic, 155
  rigid rod, 109
  ring-opening metathesis, 67
  telechelic, 25
polymerization
  emulsion, 17
  oxidative, 47
poly(methyl methacrylate), 17
poly(phenylquinoxaline), 37
poly(propylene oxide), 37
polypyrrole, 47, 133
polysilazane, 155
poly(silicic acid) esters, 67
polysiloxane, 155
polystyrene, 59
polythiophene, 47
polyvinylchloride, 47
poly(vinylmethylether), 59
positron annihilation lifetime spectroscopy, 53
preceramic polymers, 155
pressure pyrolysis, 167
printed circuit board, 47
properties
  magnetic, 177
  mechanical, 77
pyrolysis, 155
  pressure, 167
pyrolytic conversion, 155

quantum dots, 121

reactive ion etching, 47
reinforcement, 77, 91
resin
  epoxy, 11, 25, 31
  silicone, 31
rigid rod polymer, 109
ring-opening metathesis polymer, 67
rubber, 11
  -toughened, 17
rupture, 77

SANS, 59, 85
saturation magnetization, 167
SAXS, 37
scanning electron microscopy (SEM), 3, 11
scattering
  dynamic light, 109
  small angle
    neutron, 59, 85
    x-ray, 37
SEM, 3, 11
semiconducting, 121
shear, 77
shock compaction, 149
silica, 77, 85, 91, 109
silicone, 91, 115
  rubber, 91
siloxane, 59, 77, 85, 91, 115
silsesquioxane, 115
sintering
  microwave, 149
  plasma, 149
sol-gel, 67, 77, 85, 91, 121
specific fracture work, 3
spectroscopy
  impedance, 133
  positron annihilation lifetime, 53
static fracture, 3

telechelic polymer, 25
TEM, 3, 11, 37
temperature, high, 37
tensile testing, 17
tetraethoxysilane (TEOS), 77
thermotropic, 53
torsion, 77
toughened, 11, 17
toughening, 3, 25, 31
transmission electron microscopy (TEM), 3, 11, 37

ultimate properties, 77

vanadium oxide, 133
vinylferrocene, 103

xerogel, 67

zirconia, 121, 149